Just A Thought

Peter Borthwick

Grosvenor House
Publishing Limited

All rights reserved
Copyright © Peter Borthwick, 2023

The right of Peter Borthwick to be identified as the author of this
work has been asserted in accordance with Section 78
of the Copyright, Designs and Patents Act 1988

The book cover is copyright to Peter Borthwick

This book is published by
Grosvenor House Publishing Ltd
Link House
140 The Broadway, Tolworth, Surrey, KT6 7HT.
www.grosvenorhousepublishing.co.uk

This book is sold subject to the conditions that it shall not, by way of
trade or otherwise, be lent, resold, hired out or otherwise circulated
without the author's or publisher's prior consent in any form of
binding or cover other than that in which it is published and
without a similar condition including this condition being
imposed on the subsequent purchaser.

A CIP record for this book
is available from the British Library

ISBN 978-1-80381-533-6
eBook ISBN 978-1-80381-603-6

The intent of the author is only to offer information of a
general nature to help you in your quest for educational,
emotional and spiritual wellbeing. In the event that you use any
of the information in this book for yourself, which is your
constitutional right, the author and publisher assume
no responsibility for your action

DEDICATED TO MY DAD
IN LOVING MEMORY.

INDEX

ACKNOWLEDGEMENTS vii

PREFACE ix

Chapter 1: A Little Bit about Me 1

Chapter 2: Technological Boom 17

Chapter 3: My Issues with History 29

Chapter 4: Are we Alone? 59

Chapter 5: The Blue Marble 79

Chapter 6: The Modern World 88

Chapter 7: Tidying Up 112

EPILOGUE 126

ACKNOWLEDGEMENTS

My endless love and my thanks to my Mum, my children and my family for their continued support for what I am trying to achieve and for their continued love, it means more to me than I can ever convey.

A huge thanks and love to my Cousin Sarah for proofreading this book for me.

PREFACE

So, what to write here? I guess start with me. My name is Peter, and I am 51 years old, a father of two adult children who I raised as a single Dad. I have been a builder most of my life. I love a good laugh and try and enjoy life as much as possible. I am very in touch with current affairs, the world around me and nature, and love nothing better than a walk in the country or a ride out on my bike. I love a good party or night out now and then, much the same as everyone else, and like me I am sure there are lots of other people who have their story to tell. I just think mine is slightly different - so much so that I want to share my insights into, well, just about anything considered off the beaten track and outside of mainstream education and news.

My life has been very complicated for me to try and work out, trying to navigate the paths presented to us all, the good and bad aspects of life and humanity. That said, no matter what I have done in this life, I have always had something nagging at me that I was meant to be on another path, that my purpose was never what I was doing at that time, and for sure I have been pushed and forced to go into other directions my whole life, inevitably leading to this point now.

I am a spiritual man. I have deep empathy for everyone and have the ability to mix comfortably in all walks of life. I have worked spiritually at various times in my life whilst never truly knowing what this nagging feeling was that was residing within me, and knowing at each turn, however much I tried to bury it, that I was on the wrong path still.

I was expelled from school, having never taken it seriously and knew that even at that young age it was not going to be needed in the life ahead of me, only then to go into a management role on BR as it was back then (British Rail), at the tender age of 22. I was promoted yet again at 26, and once I had left BR, I went on to be a Construction Site Manager in London, followed by holding the post of Estates Manager in one of the biggest schooling complexes in the UK. I have written two books prior to this one, have built up my own small successful business over the last 15 years, and feel I could make a great case for people to stop listening to other people who run on a doctrine of thinking they know what is best for you, which is generally not a personal assessment but a generalisation covering multitudes of people, what you must do and not do, and frankly start listening to the gift we were all given - gut instinct and intuition - because in my case, even though so young when I started to recognise it, I could not have been more correct. Trust me people, we have been conditioned to live in fear of the unknown, but there is nothing to fear.

Who do you trust in this world more than anyone? That's right, you, yet most don't listen to themselves, instead following blindly a doctrine they don't really understand only to later regret not following their gut instinct.

I have witnessed the best and worst of humanity, the worst being greed, vanity, selfishness, ego, lies, and cheating. They, plus many more human failings, have all affected my life at some stage. That said, I can now recognise they were all part of the journey. I also recognise that what we are doing today, and today's stability that we enjoy, might very well be gone tomorrow - one day a homeowner with family and stability, the next day literally on your parents' lounge floor with nothing. One day enjoying a senior manager's position thinking you have stability for life, the next caught up in a web of lies and deceit that cost you your job.

These are just a couple of huge things that have impacted my life, but as I said, I feel most strongly that whenever I have settled into what I thought was my future and where I am meant to be, that nagging doubt has been there and in most cases, through no fault of my own, I have been manoeuvred in a totally different direction leading me to this point where I feel I am about to embark on another huge twist in my life story. Only this time I recognise it, as I am totally committed to following my gut instinct and understanding what I have repeatedly been told my whole life about my future. Only time will tell, I guess.

Within these pages you will find my thoughts on many subjects, some very diverse, some considered rubbish by some, some considered paranormal. You will also find my take on the universe, life, the state of the world right now, the lies regarding our history, and many things that I hope will leave you pondering it all and changing your outlook on life when you finally finish this book.

This book will follow a spiritual tone at times, but I am guessing nothing like any spiritual book you have read before. I will share some of my spiritual experiences, but not attempt to fill an entire book with endless stories about my spiritual journey. Likewise, this book will not be about developing your spiritual gifts - if that is what you seek then I humbly suggest you find yourself a suitable development circle. Among the many circles I have sat in, I once sat in the same circle, twice weekly, for over 5 Years. It is a commitment.

What you will find in here is the truth as I see it, having spent a life educating myself and trying to find the answers to life's mysteries instead of just taking what we are told on mainstream media and within education, social media, and any other biased, politically driven platform there is out there, as truth. And it's not that hard to filter your way through this web of deceit. The information and facts are right there in front of us. We have spent far too long blinded by bullshit frankly, and it is time to wake up and see the world for what it is.

Now it would be easy to put this book down now and cast aspersions my way as not another tinfoil hat conspiracy person, and I can see why you would feel that way. But please, do not make any knee jerk reactions. I can only assure you this is not the case. I look at every angle and every possibility before drawing my conclusions and try to be impartial. Investigating the truth is a passion and being a person that works with spiritual guidance at all times, where the background and foundations are bolstered up on believing in love, light and the truth, I humbly put across to you, the reader, that I am about as far removed from a tinfoil hat conspiracy theory as possible. I am sure if you met me, your first impression would be a builder-type geezer, which would be right. Well, if you just judged a book by its cover that is. Rather than trying to convince you of my beliefs and credibility as a person, let me feed you these few recent and easy to acknowledge nuggets about lies, truth and the perception of reality, before you think about putting this book back on the shelf.

For over 40 years the US Government strongly denied the existence of Area 51, yet for over a decade now we have all been aware of its existence. It is common knowledge now, yet nobody has taken time to hold the government to account for this lie. Likewise, who paid for it to be built? Who paid to run it for decades if it did not officially exist? Taxpayers money? And more importantly, how many people were ridiculed and mocked for believing it existed prior to disclosure?

When Galileo (1564-1642) announced his support for Copernicus in stating that he thought the world rotated around the sun, as opposed to the previous belief that the sun revolved around the earth, he was deemed heretical, and it was met with the upmost opposition from other astronomers and the Catholic Church. Yet, he was proven to be correct and is now revered as a historical figure. What we perceive as farfetched and beyond belief today may well be proven right in the future. It is just the perception of our reality that I will go into more detail in this book.

Not forgetting, of course, that we once thought the Earth was flat! That was mankind's reality at that time. Now, that reality has changed dramatically knowing it is in fact, round.

Let me give you this last and very recent example.

Despite overwhelming evidence all through history, people still believed that life from other planets, and life from other dimensions that resided on another vibration, were just for the tinfoil hat conspiracy theorists and the attention seekers. Yet in this last decade the US government (also France and Iran, and I have no doubt many other as yet undisclosed countries) have confirmed that they have indeed been financing secret departments who have spent millions of dollars over many decades investigating this phenomena. They have now released filmed evidence of a Tic Tac Unidentified Arial Phenomena, the new terminology for UFO, into mainstream news; allowed countless accounts from radar operators, fighter pilots, commercial airline pilots and many eyewitnesses to be circulated and released into the public domain; and the US government has at long last confirmed that UAPs exist. No ambiguity, they are real. No confirmation on alien life yet, but how long until that is also disclosed? Someone must be controlling these highly advanced UAPs!

In a nutshell, it is all about our own perception on what life is. 50 years ago, our belief was that Area 51 was a myth and not real. Now we know it is real and does indeed exist - that is a change in our knowledge which affects how we perceive life. We once all thought the earth was flat and now we know that to be wrong and it is in fact round. The entire world, including the whole Catholic Church Movement, once thought we were the centre of the universe and the sun, and all celestial bodies orbited around us. Now, we know that we do in fact orbit the sun and further to this our entire solar system is a small insignificant system at the far edge of our milky way galaxy, which in itself is just a standard galaxy among the many billions of other galaxies we have found in the thus far infinite void of space. Hardly the centre of all life, are we?

This could be deemed a huge global change in our perception of what life is and our reality, all thanks to the bravery of one very clever insightful man, who was ridiculed and deemed a heretic at the time - the equivalent of a modern-day tinfoil hat brigade man, I guess. Yet, he was right.

So many people were publicly ridiculed for stating they had witnessed a UFO as they were once called, losing their jobs and credibility, and had their lives and the lives of their family members threatened by their own governments' secret departments if they ever spoke of their experience. And now, governments are repackaging their views on the existence of UAPs and confirm publicly that they exist and that they have spent millions of dollars investigating this for decades.

So please, don't be dismissive of my insights. It strikes me that the biggest liars in history are global organisations that have indoctrinated us into following their agenda. Governments and religious movements etc, are very guilty of this farcical behaviour.

Hope fully this small introduction has left you wanting to read on further. Obviously, the tone I have set in this preface will not be the tone throughout this book for as I have said, many subjects will be covered and are not restricted to these few subject matters I have covered here. The examples given just seemed to be easy examples, as well as recent, to show you, the reader, that what we think is factual one day can change in an instant, and your perception of what is your reality can change completely in the next instance. So, keep an open mind as you read on. Some of what I write about may seem farfetched and beyond belief, but as proven all through history by forward-thinkers who looked beyond the obvious and mainstream, just because it does appear unimaginable right now, it does not mean I am wrong.

I humbly invite you to read on...

Chapter 1: A Little Bit about Me

Where to begin? I guess where life started. I was born into a single parent environment to what turned out to be a wonderful loving mum. For the first few years of my life, it was just me and mum living in our flat. My biological father died when I was 11 months old. Like all people, I have memories of those early years. I still remember my first day at school, having a photo taken outside the flat in my new uniform, the trepidation I felt about going to school. I remember my go-kart being stolen, and being allowed to stay up and watch King Kong, a repeat of the original back then of course, and as would be expected at that age, I fell asleep before it started! I recall the gutted feeling I had the next day when I realised I had missed it. No recording and re-runs back then. In fact, if memory serves me right, there were only three channels, not even channel 4 had started, so I was a very distraught boy the next morning. But mum being mum, she got me a King Kong comic the next day to console my obvious sadness at missing my first monster movie!

So, just a normal boy of that age, in a normal household. I also seem to remember there not being many cars in the road, but I guess there just were not as many back in the early seventies. So, as you can imagine, I had a very normal, loving upbringing with just mum and me to start with, nothing out of the ordinary until it was time for bed.

Looking back now, I guess what happened at night was the first indication that I might be one of those people who was very open to spiritual contact. For many years I would question whether it actually happened at all, but always and very quickly drew the same conclusion, of course it did. But that would then lead to the next question in my head, what the hell was all that about? And although I know what happened now and the finer details of those experiences, I still do not know why they occurred. Some things in life remain a mystery, I guess.

As I recall, and these memories are very prominent and as real to me as any other of my life experiences, I would go to bed the same as anybody else would. I had no fear or concern for what I knew would happen. I guess at that age, I just took it that it was a part of life and everyone went through it. I don't recall it being a concern or worry for me, quite the opposite. Once asleep, I assume I must have been asleep, I would slowly rise out of my body. I know, this all sounds like some sort of science fiction movie, but I assure you this is how it happened. I can still even now after all these years, recall the sensation I felt as it happened, even though I can no longer do this, or if I do, I no longer have the capacity to remember.

Once I had risen up and out, I would find myself floating roughly five feet above the floor, looking down at myself asleep. From here I would float effortlessly out of my bedroom and down the hall to the front door where I would be able to go no further. At the front door, there would be, for want of a better description, an entity waiting for me. It had no form as we would know it, but of course I do not know what I would have looked like when not in my human body. Maybe I looked the same. I can only describe this entity as a kind of round energy blob that blocked the door out.

I later found out through a very high-end medium when I was having a reading done, that it was one of my guides stopping me from going too far. In fact, his exact words were, to stop me going too far and seeing the dark. But not wanting to scare you reading when only on chapter one, I will touch on what I know about that later. So, to continue, I believe it was my spirit guide producing a spiritual barrier, I am not sure. All I know is that this happened to me every night for a long time.

I have no idea how long I would remain there, it could have been seconds, minutes or hours, but I can very much recall the feeling of utter contentment and warmth whilst in the presence of this other entity, a feeling like no other, something that has never replicated itself while in my body for sure. I was never given instruction to return to my body, it just kind of happened. Neither did we speak in the conventional manner. We never spoke a word and at some stage, I would float back to my body.

The interesting part about that is what I read in a newspaper article taken direct from a bestselling book by Dr Eben Alexander called *The Map of Heaven,* which I now own. First published on 7th October 2014, he was a famous neurosurgeon who had his own story to tell while going through a near death experience. I became aware of this book when a two-page centre spread article was written in the Daily Mail newspaper on 18th October 2014, this all happening around the exact same time as I myself had self-published my first book in which I made reference to the experiences I had been through as a child.

What really caught my eye in the newspaper article were the words *"Without actually speaking, she let me know that I was loved and cared for beyond measure and that the universe was vaster, better, and a more beautiful place than I could ever imagine".* What struck me is that

I had made reference in my book to much the same experience writing, *"I was never given instruction to return to my body, neither did we communicate by speaking, we never spoke a word and off I would float back to my body".* We both referenced that we never spoke a word but still communicated somehow. That really stood out for me. Here was someone who had an experience, all be it different, yet still a huge part of that experience was exactly the same - how he communicated. For me, it gave even more validation to what I went through.

Please understand that Dr Alexander prior to his coma and near-death experience actively stated that this kind of thing was not real. There is nothing after death and he thought anyone who went through this was just dreaming and it was just brain activity. Dr Alexander's story contains so much more than I will share in my book, and his vast experience is breath-taking, and all happened while he was in a coma and registered brain-dead by his colleagues. He literally demonstrated no brain activity during the time he went through his experience, ruling out his own previous assessment of what happens in these circumstances. I would humbly recommend anyone to read his book. There is so much more to Dr Alexander's experience: where he went to, who accompanied him, and further proof upon his awakening. An incredible journey that has completely changed the doctor's opinion leading him to write the book called The Map of Heaven.

Thinking back to my experiences, I can remember lowering back into my body, and I felt like I would always wake up just after this. That is how it felt, but of course it might not have been straight away. However, I do believe it was, as when I woke, I would still feel that incredibly warm spiritual glow around me at the instant I woke up.

I have no idea how old I was when this activity stopped for me, I know it never happened in the house we moved to when I was 5, so for sure it was all before that age and restricted to when I lived in the flat, but it is possible it stopped before we left there, maybe as I got older, or just stopped the day we moved. I am inclined to think it stopped before we left for the new home as I feel it was something I grew out of or spiritually learnt to control, as opposed to happening just due to the location of where I lived.

I have never properly touched on the subject with my mum, although she is now aware of what was happening back then as I have told her in brief, and I have no recollection of anything else happening at that time that would be considered different. She has said I never showed any concern about it back then or even spoken about it.

At some stage during that time, my mum met the man who was to become my dad. Sadly, he died on September 27th 2012. A few years after they met, a move for me to a new house and two new younger brothers, all that happened back at the flat at night became a memory, and something I guess as a child I just forgot about. My strange night-time activities were over. That was up until one night in the new house that I will never forget. I do not know how old I was, but I was still sharing a big bedroom with my next younger brother before dad partitioned the room to make us a room each. We had heavy-set, navy-blue curtains up and lived on the end terrace near the road, so there was a small amount of streetlamp illumination afforded to the room, but only slight as we had these heavy dark curtains.

I am a little vague about how the circumstances of this situation arose, I just recall a man standing at the end of my bed looking at me. I can't remember if my brother was awake, although he would have been very young, or if in fact I woke up to this. He was just there, he never spoke to me or made any attempt to communicate, not even a smile, he just stood there, as a child it was very frightening.

I remember my dad coming into the room, but sadly he just could not see what I could. He was a decorated war veteran and fought in the Korean War and was not easily shaken. I vividly remember him playing with the curtains and trying to convince me that it was just the streetlamp outside. In fact, he walked through the man on a couple of occasions whilst trying to console me! Imagine how I felt about that! A most unnatural thing to happen right in front of my eyes! My dad did his best, as any good dad would, to alleviate my concerns and if I remember rightly, I just hid under my covers. I can still see the man standing there now. It will never leave me, although I am confident that he meant me no harm, and I am still unsure who he was. As a child I was naturally just scared!

This all happened before my tenth birthday and was to be last the last I saw spirit with my eyes for some time. As I headed into my teens, life became very difficult and odd for me.

Early school life for me was nothing out of the ordinary. During my time in infant and junior schools I was content just to toe the line and don't recall anything strange about school life back then. I am not vain enough to say I was super popular, but I have never struggled to make friends, and had as I recall at that time, a good childhood relationship with many other kids, both boys and girls. I very much enjoyed junior school. That said...

My time in senior school concluded with me being expelled, having never really felt like I should be there.

It is hard for me to explain exactly what senior school was like, and I would love to meet with someone who can relate to my emotional state at that time. I was never properly happy at school. To any teacher I am sure I was a nuisance to say the least. I still have most of my school reports, and they make humorous reading now. The stand out comment was *"Peter gains very little from school and contributes even less"*. That was the sum total of what one teacher said about me for a whole year. For some this would be a damning verdict, but for me so much time has now passed and with my own life experiences since, I just chuckle at my school reports.

That process (Schooling) has had little relevance in my life having never once been asked to present evidence of my school qualifications in adult life.

I could be in a minority off course.

I went to an all-boys school that had only changed in my first year there from grammar to comprehensive. It was 1982 and the school still worked within those old grammar school traditions and rules. In my first couple of years there I remember some staff members still wearing mortar boards and gowns or (squire cap's and capes as we called them). In the modern world, and reflecting through eyes that live in 2022, I am not entirely sure what the purpose was for dressing up like this, it would be hard to make a case that it improved the level of teaching given simply by wearing this attire, and if it was done for intimidation then sadly in my case it did not work.

I guess this kind of attire was just of the time and a personal choice for the individual teacher as some of the teaching staff were comfortable in jeans and t-shirt. There was no consistency at all.

I say all this in understanding and appreciating very much that graduating students look great in their mortar boards and gowns as a part of that celebratory moment, but as a day to day working dress-code the concept is lost on me.

I think it was during my school days when I was coming to terms with my extrasensory perception. As said, I could feel people's emotional state, and imagine having that in a place full of teenage boys, including many teaching staff that I felt were clearly unbalanced as well. We are all human and teachers are no more unaffected by life's trials than anyone else, and life outside the classroom can unfortunately affect life in the classroom for both student and teacher alike.

Obviously there are many situations that as adults we go through that affect our emotional state and by extension, our energy field and aura, and as with anyone working in any other chosen profession or industry, teachers are no different. They like us all, are affected by daily life, we are all in the human experience after all, nobody is exempt, and for me and some of my class mates, we witnessed that some teachers behaviour at that time, frankly meant they should not have been in that environment at all. For me with this developing gift, it was more than the eye could see.

I could feel all these different emotions and out-of-balance people, but unfortunately, I was too young to understand most of what I was feeling to even begin to show empathy towards those around me and frankly I was trying to deal with my own understanding of myself. To me, I was alone in whatever I was going through, and escape was the easiest answer which required bad behaviour to achieve that goal.

I think I just could not control it, which in some way led me to act the joker, make my classmates laugh, happy laughing people was a better emotion to deal with.

Inevitably I would get in trouble; it was all a-part of my escapism, without being consciously aware of why I was behaving like I was.

I am not sure that in the four and a half years I spent at senior school that I ever produced anything worth looking at. My reports really do have nothing good to say about me, although it is funny stuff to read now. I think the only time that I had anything written about me that was positive at that time in my life, was when I did my work experience week, a glowing report, but that was in a totally different environment that probably suited my free spirit better.

Back to school life and I recall flooding the senior school lecture theatre, setting off the fire alarms to get out of the next class, writing obscenities everywhere. I regularly climbed out of the classroom window while the teacher wrote on the black board, much to the amusement of my Classmates.

I generally turned up late every day, sometimes went home at lunchtime and just never went back, was always over the back of the sports fields smoking. I wore a denim jacket to and from school as opposed to the blazer and had Bill Parker written on the top left pocket in marker pen where the school badge would be on the blazer, with the full name of the school and not the abbreviated name that I had written. I must have been a sight walking to school about ten o'clock in my denim jacket smoking a cigarette with no books to my name, it did not help that while at senior school we endured a very long period of teacher strikes where we were sent home at lunchtime. I just could not adjust back to normal school hours when they were over; I liked the new hours and stuck with it as much as I could.

I think my sole goal back then was to think of the next silly thing I could do without getting caught that my friends would laugh at, and would distract me from emotions that I felt. I just did not realise at that time I was different. Now I can forgive myself for behaving like that and understand why I was so silly, sitting in a classroom listening to someone drone on, which was no fault of the teacher who may well have been exceptional at their job, but if you feel like you just do not belong there, like I did, it made you feel like this is just someone droning on, it was not for me.

There was fun to be had and I spent a lot of time looking out of the window wishing I was outside in the world and not trapped in a prisoner environment where I felt I was gaining nothing. Even then, I had an in-built sense of knowing that I would get information from a far higher and more divine source than school when I would need it most.

I also look back and think about what this must have been like for not only my teachers having to try and work me out daily, but also my poor parents, getting a phone call every day from my form teacher telling mum what I had been up to that day, I guess to try and garner some support from them.

Regarding my erratic behaviour, my laughable school reports, parents evening must have been so embarrassing for two people who were very good conscientious parents. Please understand that I was never a nasty lad and was very kind-hearted. My 3rd year form teacher made note on his report that I was very kind hearted and meant well, which probably made me more of an enigma offset against my rebellious attitude.

All that said and with the beauty of karma, a couple of years back, while working, I was in a small paper shop on the outskirts of town buying a sandwich for my lunch when a man stood next to me doing the same thing.

I recognised him after about 20 seconds of curiosity, it was my old form teacher from my 4th and 5th years of senior school, a man who I felt demonstrated regularly that he hated me, probably for good reason, and someone who looking back I did not like at all based on my memories of him. He would regularly call my mum to inform her of whatever terrible thing I had done at school that day, and I felt it was to try and get me into trouble at home, only to be further mystified when my mum would inform him she already knew as I had told her, I never kept much away from my parents about what I did, or in most cases, what I did not do at school.

I felt back then at school and standing in this shop that our mutual dislike for each other was more than tangible, it was clear and present, boarder line hatred. That said I was no longer a teenage boy and would never dream of approaching any situation in any way other than positive, my curiosity compelled me to introduce myself to him.

So, I asked this man if he was who I suspected and he confirmed he was, I introduced myself and he took a few seconds to absorb the information that this middle-aged man with a greying beard in front of him, politely asking his name, was in fact the terror of what was probably his early years of teaching.

I asked how he was, and he told me that he had recently lost his wife, which I expressed my condolences for and he asked me about my life, which I briefly covered, mostly about my grown-up children and how well they were doing in life, my work etc. Then he mentioned that amongst his recent woes he had broken his neck a few years back, C6 and C7 were damaged. This shocked me as I too had broken my neck at C6 and C7 some years back, which gave us a mutual subject to chat further about and bond in some way, even discussing our hospital experience, as we were both admitted to the same hospital to receive treatment.

After a while it became apparent to me that anyone else watching us would think we were just two old mates catching up, not a teacher/ student relationship that I felt was based on pure loathing of each other.

When we eventually exhausted our conversation, we shook hands and said how nice it was to see each other and catch up and I left feeling like some form of karma had taken place for me and possibly us both. How nice for us both to get the chance to put all demons to bed. How wrong we could both have been about each other, and all the animosity I felt we had towards each other some 30 plus years ago, got wiped away in an instant, all gone.

I genuinely feel so happy to have met the real man, and I like to think he left feeling the same way. He certainly seemed to be genuinely pleased to see me and whenever I think about that situation, it brings a smile to my face now.

How many of us are quick to judge others, harbour lifelong hatred, when really when you get down to basics, we are all much the same, the same problems and same issues? In this case you had two people, one an adult, one an adolescent child, one a teacher, one a student, regularly demonstrating dislike towards each other, making each other's lives a misery each day, basically when it comes down to it not understanding the other, only to find out that if put in different circumstances we were closer to getting along than we ever imagined.

I look back at that time in life now understanding why I was the way I was. I am now at ease with that time in my life, have forgiven myself and asked for forgiveness for any miss-doings I did that hurt anyone, mentally or physically. I can now deal with the emotions that were manifesting back then so much better now. I do still have to deal with the abundant amount of empathy I have for everyone, and sometimes I am overwhelmed with emotions that are not mine, but I have learnt to control it better now.

As with us all, I have had many things happen to me, witnessed many things and heard so much that has had profound impact on my life. Some of it was truly horrible and stripped me physically, mentally, and materially right back to my bare bones, and took me years to build myself back up on all those fronts. Some things were far more subtle but being a man that listens and observes life's subtle signs, I have been able to identify and acknowledge when these subtle but significant occurrences happen and keep them with me to refer to later in life when they become relevant again.

One of the many significant times in my life was when my work colleague and friend was killed in a lorry accident. Shocked and upset would be an understatement of how I reacted at the time but having been introduced to mediumship and spirituality at a young age by my mother, I sought out a respected medium to see if any sign of my friend could be found.

I found a local but respected medium who worked by drawing what he received - basically he drew a circle and filled it with many smaller drawings which were the basis of his reading. I had to ask at times for him to confirm what he had drawn, and if I am honest most of it made no sense at that time.

Now that so much time has passed however, I can see most of that reading was based on future events. Please do not confuse a medium with a fortune teller, they are not the same thing and I have no validation or experience to fall back on to vouch one way or the other for the credibility of a fortune teller.

So, the stand-out things that were drawn at that reading were a house on stilts with a bed and pound sign within it.

Upon asking him what this meant for me, he stated that my current relationship was built on dodgy foundations and that we would separate due to money issues and the bedroom. This was one month before I was due to be married. We eventually split 5 years later due to repeated adultery on the part of my wife where upon she tried to take the bulk of the equity from the house sale too. None of this was on the horizon before being married. We had been together many years prior to being married and had a child. Either way, the reading was bang on correct!

Other small things on that reading were also correct. A small wishing well was drawn, not a very common thing even in the UK. I took a job many years later in a private school and there was a wishing well just outside my office.

It was during my time in that job that I became involved in an employment tribune for unfair dismissal. There was reference made to this within that circle and the fact that I would be successful, including the name of a work colleague who worked very closely to me. Her role was admin support to my management position, and she stood by me throughout the process in understanding I was wronged and wanted the truth to be aired. As said, her actual name is there in writing right next to the legal references, in this case pictured as a judge's wig, with the words success and settlement.

I kept to my guns and a payoff of a few hundred pounds I was offered at the start, increased to tens of thousands of pounds after a lengthy battle to prove my innocence. I had not even heard of this school at the time of the medium's reading, and along with the wishing well, I would say he was bang on again.

The reading also contained many past life regressions going back many thousands of years for me. Again, I asked about this and was told that my soul is very old and this is why I find looking beyond the mainstream into things considered taboo, like mediumship for one thing, and uncloaking all the nonsense we are fed, very easy to understand. It is due to the fact that my soul, for want of nothing else to call it, has been around for a very long time and has experienced so many lives on the earth throughout history.

A stand-out piece relating to past life regression was that I was in charge of the Archives in Tibet in 2000BC. The medium told me that my soul was already elevated to a level that I could come to the Earth plane and be placed in this responsible position, even though it was 4000 years ago, and I must have already been an old soul even back then, so long ago. This really did start me to question who I was and what I am doing here this time. I understand this all sounds like a fictional story, but these were his words in describing what he was drawing.

Beyond all this though was what we discussed when the reading was over. I was given the opportunity to ask a question at the end, and completely off subject I asked if UFOs were real, not aliens, UFOs. It was a complete knee jerk question that popped into my head, to which he said yes, they are real. He then pondered for a brief time and said that it will all be disclosed in my lifetime, meaning me (I was much younger than the medium, 26 at the time).

I do not believe we have full disclosure yet as nobody has confirmed alien life yet, despite overwhelming evidence, but we have had disclosure on the fact that UFOs, or UAPs, are in fact real. This is no longer a secret. So, he was right, and again this reading took place in 1997 when that information was strictly off limits and taboo and reserved for the conspiracy theorists.

Lastly, as I was leaving and actually going out of the front door, he stopped me by placing a hand on my shoulder and told me that I would work with my friend again someday in ways I could not imagine right then. That has yet to happen, but I feel I am on the cusp of that statement becoming reality for me. My friend would come through and contact me through other mediums I had readings with many years later, as well as family members that have passed over, including my biological father.

Anyway, that is a little bit about me to help you to form a picture about who I am. I am sure as I write this book other things regarding me will be written but for now, I just wanted to share some of me and some of my life experiences and experiences with spirituality.

Chapter 2: Technological Boom

As well as questioning things and looking beyond what the mainstream offers, I am also one of those people that ponders things in hindsight. Why has this happened? Why are we here? etc, and one of my most compelling ponders has been the current technological boom we are living in.

When you look back through history, our progression has been a methodically steady-paced evolution, Advancements in not only how we conduct ourselves as a race and interact, but also our advancements in technology, up until the last 70-80 years, have been reasonably standardised and paced. Whereas nowadays, our current technology is outdated as soon as it is released. Based on our historical progression which can be traced back thousands of years, this indicates that we are currently involved in a huge leap forward in technology that is completely outside of our historically traceable evolutionary path.

When pondering this subject I try to look at every aspect, and some would argue and make very good cases that we as humans have just managed to unlock technology at a rapid pace and as one advancement presents itself it helps and leads to the next, as proven with our constant upgrade lifestyles. Gone are the days of replacing something when its broken, we now upgrade at an alarming rate given the lack of resources on this planet. This is something for another day I feel but it does indicate how quickly technology moves these days.

Some would argue we are just smarter than we used to be. Well, define smarter. Someone who is considered very smart on the African savannah, who for an example of his skill set can predict weather forecasts just by studying the sky and surrounding winds, would be rendered near useless in the New York stock exchange.

Since IQ tests began about 100 years ago, we have been steadily getting higher results, an average of 3 extra points per decade, but as stated, intelligence is all relative to which field you work in and what life you live.

A neurosurgeon with a huge IQ would be considered pretty far down the IQ food chain on a building site surrounded by construction tradesmen.

Granted the neurosurgeon would probably have demonstrated a higher IQ result than the general populace on the building site. I feel that is a fair statement and I am a builder. But taken out of his field of expertise he is rendered almost useless and could be seen by others in the building site environment as not a very smart man at all.

So having pondered this subject for some time I have looked at the possibility that this technological boom was kick started by other means as opposed to humans unlocking previously undiscovered technologies through the natural progression of human skills.

What I propose to write about in this chapter is the possibility of an alternative to current thinking regarding our boom.

I hope to make a compelling case to at least challenge the idea that we unlocked all this technology ourselves. Don't get me wrong, some credit must go to man's indomitable quest to discover new things, but given the pace and timing of this technological boom my insights have led me to believe it kick-started somewhere other than at the hand of mankind.

In this chapter, I intend to give some examples of where we were as a race just a hand full of decades ago, where we are now, and my thoughts on what possibly kick-started our current tech boom.

Now it is not my agenda to make this book about alien life, although we do have to acknowledge the huge impact that these entities, phenomena, life-forms, whatever you wish to call them, have had on us all though our history. These beings have been visiting us for thousands of years, and the evidence for otherworldly beings playing their part in our history is evident and tangible so I must now include a small part of that subject matter in this chapter.

Given we have now had disclosure on UAPs, there must be a chapter I can write in this book? To further cement the case for alternative life forms.

For now I wish to allude to one particular incident that corresponds with the beginnings of our recent technological boom and where it took us. That incident happened in Roswell New Mexico in 1947.

I have to assume that you may not be aware of the events of Roswell as it has come to be known as, and so let me briefly outline the events.

Unfortunately the Roswell incident has been tarnished and made into an almost comical event, but the true facts are actually startling.

On July 8th 1947, Roswell Army Air Field issued a press release, probably in all innocence and in pure honesty. Governmental cover-ups were not in practice at this time, stating that they had recovered a "Flying Disc".

This press release was carried out by the Public Information Officer Lieutenant Walter Haut under the authorisation of the then base commander Colonel William Blanchard.

Major Marcel was placed in charge of the recovery of the debris.

24 hours later and under instruction from General Roger Ramey a further statement was released stating that the debris was in fact just a downed weather balloon.

Please understand that a weather balloon is not a complicated piece of machinery. They are very basic in construction, and Roswell Army Air Force base staff were considered the absolute elite, very intelligent members of the armed forces. Roswell is where the Hiroshima and Nagasaki atomic bombings were organised and were undertaken by Roswell staff, considered the best of the best.

Yet we are to believe that the staff on the ground and senior commissioned officers like Major Marcel and even the base commander Colonel Blanchard could not visually identify something as simple as a downed weather balloon mistaking it for a crashed disc, something constructed of tin foil and scotch tape.

The debris was also witnessed by multiple people including everyday civilians as being taken away in sheeted trucks with armed escorts, slight overkill for a downed weather balloon?

Through the rest of Major Marcel's life and career he was obliged under orders I would imagine to stand by the weather balloon story.

Much later in life he has been noted as making a statement, describing the recovered debris as "nothing made on this earth" and that the material was very thin but very tough.

Rumours also circulated about the fact that aliens bodies were also recovered, including one that was still alive but died shortly after the recovery took place.

Lieutenant Walter Haut, the Public Information Officer who originally released the press statement left an affidavit to be opened only upon his death.

That statement asserts that the weather balloon claim was a cover story, and that the real object had been recovered by the military and stored in a hanger. He described seeing not just the craft, but the alien bodies too.

Many other people involved with the Roswell incident have also come forward over the years to confirm the cover up story, hands on soldiers on the ground etc, but here I have just made note of the higher-ranking officers who were at the heart of the press release, recovering the debris, change of story, cover up, and later the revelations of the truth.

Shortly after the events of Roswell, technology took off at a rapid pace in both civilian and military applications, was this us undertaking a reverse technology procedure? New Technological developments took us on the battle field from woollen jumpers to bullet proof jackets; from bombing runs that could fall victim to the weather, high winds, and lack of moonlight; to precise day or night missile strikes over distances of thousands of miles; from prop planes that operated at an altitude no higher than 30,000 feet and at speeds no more than 400 mph, to highly advanced super-sonic planes that skirted the edge of space at speeds well in access of 2300 mph. This too became redundant when undertaking spy missions with the modern day satellites, and now off course everyone has access to Google Earth.

Civilian life has benefited too, microwaves, I-pads, mobile phones, digital cameras, internet, satellite navigation, Apps, on-line banking, home computers, none of those things just mentioned were even thought of when I was a child.

Televisions have gone from big clunky boxes only 25 years ago to lightweight hang on the wall units that are no thicker than a picture frame in depth and growing in overall size with every upgrade and all with digital picture quality, essential considerations to any lounge furnishings were the TV unit, now almost rendered redundant.

When I was a child, I had to get up and turn a switch on the wall to change the channel from the 3 analog channels available. Then comes along the remote control, in these modern times and based on my own household, this small hand held device gives me access to Free-view TV, Sky TV, Amazon Prime, Disney Channels, Netflix, and literally hundreds, possibly thousands of options and all with digital quality pictures controlled from a small hand-held device not connected to anything other than my hand.

I am only 51 as I write this, tell me at 10 years old when watching my clunky box with what turned out to be a very poor picture of what was to come, I am not sure even at that age I would have believed you.

My First home computer was a ZX81, when I was a teenager. It had a 1K memory. I am not sure what that equates too in modern terms, but I am mildly sure the mobile phone next to me probably dwarfs that kind of memory by many thousands of times, probably millions of times!

I used to have to load information on that old computer using a cassette tape that could take up to 20 minutes to load, sometimes failing to load, so you had to start that process again. Like everyone else now, I get impatient if my

JUST A THOUGHT

computer buffers for even a few seconds when accessing the world wide web!

I went to watch the Filming of a television quiz show once, and a question was thrown out to the panel and the audience. The question was, "How many memory units from the 1960s would it take to match the memory capacities of one modern day mobile phone memory card?" The host had one of the 1960s units to demonstrate, and it was about the size of a packed lunch box. The mobile phone card was roughly the size of my finger nail and a millimetre in thickness.

Many answers were thrown out by the panel, ten, twenty, anything up to fifty were offered to the host as plausible answers.

The exact answer blew me and the other people in the studio away. It would take making enough of the lunch box sized unit's to match in weight the equivalent of two modern day Air Craft Carriers, this to match the memory capacity of this tiny card the size of my finger nail; an extraordinary leap forward in our technological evolution and this was achieved in the space of fifty years.

Here are some other facts with dates that further highlight my thoughts on this matter.

The first Laser was built in 1960.
Spy Satellites were first invented in the 1960s.
The Lockheed SR-71 Blackbird took its maiden flight in 1966.
SR-N4 hovercraft, started service in 1968.
The Apollo 11 moon landing took place on 16th July 1969.
Reliable computer networking 1960s-1970s.
Fibres optic cables, the groundwork and revelation of the physical properties of glass that hugely increased

high-speed data communications were found in the 1960s and perfected in the 1970s.
Missile accuracy through precise navigation 1970s.
Concorde took its maiden flight in 1976.

Kevlar, which today is used for anything from racing sails to bulletproof vests, due to its high tensile strength to weight ratio, by this measure it is five times stronger than steel, was discovered in 1965.

Mayor Marcel of the Roswell incident, July 8th 1947, stated that the debris found was not of this Earth and was very thin but very tough. 18 years later we perfect Kevlar, which if described by its appearance would be called very thin but very tough..... Coincidence?

I consider the above to easily make it on the top 100 list of the most technological advances in the last century if such a list were compiled.

What strikes me are the dates. The bulk of the most publicised and biggest leaps forward were made in the 1960s and 1970s; supersonic air travel, hovercraft of enormous proportions, landing on the moon, bullet proof jackets, precise strike missiles, the list goes on. Hey while I write this I just remembered colour TV. Guess what? Developed in the 1960s. Is it coincidence that the decades following on from the Roswell incident conjured up all these advancements?

It occurs to me that some reverse engineering took place over a period of a decade leading into the 1960s, when those applications were put into use.

Or through the 1960s that reversed technology that had been unlocked in the preceding decade, the 1950s, and was being perfected in the 1960s, was put to use in the 1970s.

Whatever the case, the three decades following on from Roswell produced some astounding advances in technology and scientific breakthroughs.

Whether standalone advancements, or in fact just progressing what we had already thought of and invented, and then taken it to a far greater and quicker evolution than we would have been capable of on our own, I cannot confirm, I cannot even confirm that my thoughts on this matter are correct, but certainly a pattern is there, and a compelling case could be made to challenge mainstream thinking and that we made these breakthroughs.

The trends continue into today, with more recent inventions like the Hubble telescope (1990), the internet and digital TVs and cameras, to name but just a few.

As a point of interest, NASA through their developments, gave us car seats that are far more compatible with the principles of ergonomics, Nissan were the first car manufacturer to use NASAs "NBP" *Neutral Body Posture* or a posture the human body naturally assumes, to develop their next generation car seat, making them far more comfortable and posture efficient.

One of the biggest misconceptions is that NASA are not part of the military as they are promoted as an independent civilian administration, well they are not and are completely funded, budgeted and I have no doubt managed and controlled by the military, the same military that recovered the debris from Roswell in 1947. NASA (National Aeronautics & Space Administration) was formed in 1958, 11 years after Roswell.

It strikes me that NASA, who would have been privy to some of that reverse engineered technology could have leaked that technology, at least the stuff they did not hold onto for military use, into the public sector, whether covertly or openly like the NBP seating.

Other technology, for use in civilian applications also followed on, things like Fibre optic cables, Kevlar etc.

Shortly after the Roswell incident the sound barrier was broken. Now it is my belief that this development was inevitable and we were well on our way to achieving this prior to Roswell, but the advancements in that field made shortly after Roswell are staggering.

By 1976, 29 years after a first-generation jet passed through the sound barrier, flights were available to civilians to travel up in the stratosphere at supersonic speed, twice the speed of sound, in luxury, with seating for up to 122 people on Concorde.

Military advances were even more astounding. The blackbird jet alluded to earlier, was introduced in 1966, 19 years after we broke the sound barrier. This jet flew on the very edges of space, some 30,000 feet above what was the Concorde altitude flying ceiling, and this done at super- sonic speeds in excess of three times the speed of sound.

Please bear in mind that we first learnt to fly in 1903, on the Kitty Hawk invented by the Wright Brothers, which was no more than a glorified kite, and by 1915 the fasted speed man had achieved was 200 mph.

Less than one lifetime later we are travelling in luxury in civilian clothing, eating dinner washed down with champagne up in the stratosphere where the curve of the Earth is visible at speeds in excess of twice the speed of sound!

66 Years after the Wright brothers first flight, we landed man on the moon. In the space of less than one lifetime, we went from a crude flight on a beach that itself only managed

a shorter distance than the wing of a modern day Boeing 747, to man walking on an entirely different world.

Today we have sent probes to the far reaches of our solar system and beyond; landed robots on other planets; through the advancements in satellite and telescope technology we have managed to map the known universe; and look at light being sent from stars that are so far away the light travelling at 183,000 miles a second is many billions of years old. I think we can all agree that is quite a head mess and a huge technological advancement!

183,000 miles a second equates to travelling to the moon and back in 2.5 seconds, travelling around the Earth 7 times in just one second. The light we are witnessing with our new space telescopes has been travelling at that speed for over three times the length of planet earth's entire existence... total head mess!

So many of the things we take for granted in our modern lives simply did not exist a few decades ago.

Not since the industrial revolution has mankind made so many advancements and even by the standards set back then, they are completely overshadowed by what we are experiencing right now. A modern day technological revolution might be how future historians see as what is happening at this moment in time.

Never in mankind's traceable history has the human race advanced so quickly; are we just benefiting and enjoying technologies that we have reverse engineered from other races and civilisations far in advance of humanity?

If the debris at Roswell was a craft from another world, and such a craft were literally hundreds maybe thousands of generations ahead of us technologically, which given the footage and eye witness accounts from highly trained fighter jet pilots about the TIC-TAC UAP they witnessed, they clearly are beyond anything we currently possess on this planet, do we think it would have been assembled using such things as copper wiring, rivets, or nuts and bolts? Highly unlikely.

Imagine if we first saw Fibre optic cables instead of inventing them, super-fast, super-efficient conduits for communication and thinner than hair. If the process for reverse engineering started back in 1947; I guess it would have seemed very off-worldly at that time. But reverse engineered and released to be used in civilian applications, it has progressed the world of communications tenfold.

Gone are the days of uncomfortable cars with radios that were tuned by turning a knob and pulling a large metal Arial out from your car bonnet area. Now cars have digital radio with the reception picked up from an Arial hidden within the car. Technology has given us in-car phone capacity, satellite navigation, voice recognition text messages and voice recognition engine ignition, light-reactive mirrors to name but a few features within the modern car. Most applications are controlled from your steering wheel all while sitting in ergonomically proficient seating, none of which was available just 20 years ago.

The list is huge and coincidentally has all been developed since an incident in Roswell, New Mexico in 1947 that in the first instance made global news as being a recovered alien disc.

I am far from saying I am totally correct, but the deeper I delve into this, the more it gives me food for thought and hopefully now you too have paused to give this subject greater scrutiny than you had prior to reading this chapter.

I intend to write a chapter later titled *"Are we alone?"*, Maybe that will add more meat to the bones regarding what I have written about this technological boom when I cover the overwhelming and tangible evidence of off-world visitations to planet earth over many thousands of years.

Chapter 3: My Issues with History

Part A: My Opening Thoughts

I guess you could say that my issues with history stem from the incredible amount of diversity there is out there regarding our origins, the origins of our planet, our many faiths, our sketchy at best understanding of ancient civilisations, and most compelling, what we have unearthed through archaeology and technology, discoveries that frankly throw some of our previous understandings and teachings right out the window.

Examples being structures that we would struggle to build even today. In fact, some are still impossible to build now yet have stood for thousands of years! Huge underwater cities that are being discovered are far in advance of what we understood about civilisation at the time they were built; skeletons of huge proportions that hint that giants, as outlined repeatedly and clearly in many historical texts, once shared this earth with us, one example being in The first book of the Hebrew Bible and the Christian Old Testament, Genesis 6:4 states, *The giants were in the earth in those days*; carbon dating of human remains that date back far further than we believed our history went; wall drawings and mosaics that either stand alone or in Egyptology sit side by side with hieroglyphics and clearly show what look to me to be some form of alien being and what must have been a giant humanoid life form; structures all around the world of very similar and more than coincidental design that were all built before the other continents were discovered, yet we were all building the exact same things at those times; structures that line up with celestial bodies and compass points to almost pinpoint accuracy, way more than just luck, before we had even begun to map the stars.

The list is endless, and with my nature and my interest, I have tried my best, understanding I am no expert in any field, to try and make sense of it all - tried to find what I believe to be the truth based on what we have discovered with tangible evidence and attempt to give you what I believe to be the truth about our history, or as in the previous chapter, at least challenge some of what we believe and have been told about our history.

Part B: Religious Explanations

So, where to begin with so much subject matter to try and wade through? I suppose with what I was told as a child regarding the creation of Earth and man. I went to a state-run junior school that still worked under the doctrine of the Church of England. As such, we sang hymns in assembly such as "He's got the whole world in His hands" and "Lord of the dance" etc. We prayed daily in our assemblies - it was all the norm back then.

This bit is difficult for me as I am working on memory from a long time ago, and it gets cloudy as to the origins of my information. I have stuck in my head two versions of our origins that may have worked together or have been separate stories. I think they came from the Old and New Testament. Either way they came from my school life and obviously came from somewhere as they are installed in my brain, so apologies if my facts are slightly out regarding what Testament these origin stories came from.

The first version that sticks in my mind was the Adam and Eve story and the Garden of Eden, which I believe came from the New Testament and was written by Christians? I cannot dismiss this straight away and will come to my thoughts on this version later in this chapter when I can expand on what I think with more clarity.

The other was what we were also taught, or certainly I was made aware of, and that came from the Old Testament, the fact that God created the world in 7 days. I am sure we are familiar with the story: on the first day God created light, on the second day God created "expanse" or as some translations call it a "Firmament", carrying on right through until the creation of man in His image and giving us dominion over the fish and over the birds and over the livestock all over the earth. Then, God decreed that the entirety of his creation was good. Quickly and off subject, given our war like nature, greed, vanity, lies, consumerism, selfishness etc, I humbly submit this decree was wholeheartedly wrong - creation was not good! If this was the true case of our creation, which I do not feel it was, things are not good, we are in fact a total failure.

My issue with all this is that through that entire process it alludes to the entirety of God's creation being set up for the benefit of not only Man but of the Earth too, and that the sun, moon and stars are all for the benefit of Man, which we now know is clearly not the case.

Given the virtually impossible-to-comprehend size of the known universe, the near infinite amount of galaxies we know of, to say nothing of the billions upon billions of stars and planets that reside inside each one of those galaxies, the now very strong possibility that other dimensions do in fact exist, I find it hard to believe this version of our beginnings. It is very primitive in its nature, belonging in the past where our attempts to understand our existence would have been hindered by what we could actually prove. It was far easier to attribute unknown information to an all-knowing and benevolent God, which historically is what we did.

To have the power and knowledge to have accomplished this task would take a being of an omnipotent and omniscient scale, and if such a being existed surely he would have had the intelligence to not have created so much for so little. It just makes no sense to me, and given that we now know that we are just a small planet orbiting a small star on the edges of our galaxy, and that through our technological boom we have discovered that every star we have managed to study so far in our milky way, not even close to 1% yet but still over 3200 in total, all have a planetary system orbiting them, it is very unlikely we are alone in this universe. In fact, mathematically the chances are zero, as is the fact that it was put together in seven days just for the benefit of mankind and planet Earth.

I have also heard the theory that we once resided on another planet and that we came here across the stars. Now that may well be nearer to the truth than you think, which I will come to later, but surely if we had the technology to have made such a huge journey, we would have arrived here far in advance of where we are now. Where did all that technology go if we are in fact space travellers?

Lastly for this part and a little off subject, but I guess as this part is titled religion it would be better slotted in here, I am moved to share with you something I've heard so much about ... I want to draw your attention to the fact we are all told that God is full of unhindered, unlimited and totally unbiased love for us all, and yet in spite of all this love, frankly life down here on Earth can be bloody difficult most of the time, and I consider myself lucky to be in a part of the world that is developed.

Given just how hard life is with things to balance such as poverty and ill health, does anyone else think that with all we have to contend with in our lives, this is a funny way of showing love? I have heard such teachings as *God wants to help his children be happy, He has blessed us with so much, He loves you and watches out for you*. Well, what about the billions who live on less than £2 a day? Not showing them much love and looking out for their best interests I feel!

I recently caught an interview with the author and star of television and film Stephen Fry, widely considered a very intelligent man. He was asked about atheism, he being an atheist himself, and to explain what he would say at the pearly gates if heaven and God did in fact exist. I will not write here word for word what he responded with but his opening statement definitely struck a chord with me and made me pause to think:

"Bone cancer in children, what is that all about? How dare you create a world that has such misery that is not our fault, it is not right, it is utterly, utterly evil, why should I respect a capricious, stupid mean minded god who creates a world that is so full of injustice and pain?"

Stephen went on to make further statements which, as said I will not write it all here, but I think just his opening statement was enough to stop me in my tracks and listen to what he had to say. I highly recommend anyone to listen to this interview.

In a world where we are repeatedly reminded of the fact that we are blessed, made in His image, loved beyond measure and all we have to do is bend a knee and pay homage to this God at church, rightly or wrongly, it was refreshing to hear someone look at it from the other side of the table and lay down some facts that contradict the religious doctrine about God's unlimited love for us all and present an alternative way of thinking.

Part C: Dinosaurs and Extinction Events

One of the key things we have discovered about our planet's history that further complicates our origins, is the discovery of the dinosaurs, who roamed this planet between about 265 to 66 million years ago, which again throws the religious explanation into disarray with humans not occupying the Earth back then as these time lines were many millions of years before the first modern humans / Homo sapiens appeared. The other interesting thing about our history are the many extinction level events that have taken place - the worst was roughly 265 million years ago, just prior to the rise of the dinosaurs, the last one was 65 million years ago that wiped the dinosaurs out. Geologists can with reasonable confidence state that there have been five extinction level events in total, the other three way back in Earth's history and prior to the worst event 265 million years ago.

My thoughts are that this seems to be a recurring theme, all be it huge time spans between each event but something that will inevitably happen again. Also, and in understanding that the Earth is approximately 4 billion years old, there could have been many races and civilisations that lived on planet Earth that achieved the same zenith of technology that we are currently enjoying, and were wiped out in an extinction level event, and that event was so long ago, billions of years as opposed to millions of years ago, that all trace of their existence has been wiped away with time, or has it? There is compelling evidence provided by geologists that during this planet's history it has already succumbed to a nuclear war, so the question is, between whom? One thing is for sure, if we are wiped out through an extinction level event or through our own stupidity and wars, the Earth will continue to spin without us, and given what we know so far about the planet Earth's past, Earth will undoubtedly spawn new life again.

Part D: Separating Planet Earth and Humans

At this point I wish to separate the beginnings of the Earth and the beginnings of man, as it simply was not at the same time.

Truth be known, I have no idea how the Earth and galaxies were created, and probably we will never truly know for sure. I have studied the Big Bang theory, and it makes sense to some degree, given that the universe is expanding at all times, it makes sense that if everything is expanding and moving away from each other then at some point in time everything was condensed into one singularity. My question for that would be two-fold. Firstly, how could so much be condensed into so little space, the known universe in one singularity? Seriously, I cannot see that. As yet we do not know the extent of the universe. It might be infinite, which squishes that theory straight away. Secondly, if this was the case, who in the first instance created the singularity that went bang?

All we can do for now is theorise and try and make sense of our beginnings, but for sure what I was led to believe in my youth about our creation has been challenged and in my head superseded by what we now know about our place in the cosmos.

Anyway, that is about as far as I want to run down the rabbit hole with the origins of the Earth and the other celestial bodies as I really don't have any answers to give. I just have, as the title of this chapter would allude too, issues with our history on the subject of the creation of the universe and everything within it, including our planet. Hopefully, I can shed more light on the origins of man, our history and how we came to be the dominant species of this planet further into this chapter.

Now, I must again emphasise that I am no expert in any particular field. My thoughts and conclusions are based purely on what I have studied for myself and through my meditations that link me to what I like to call the universal consciousness, which if I am honest, tweaks my brain into questioning so much. I will try and cover the universal consciousness later in this book.

Moving forward, I will do my best to un-fog all the varied sources of information and the clues found that form the foundation of my thoughts and conclusions and hopefully I can present these ideas in a way that is interesting to you, the reader, and can write in a way that is easy to follow and not clogged with endless scientific references and metaphors. I am passionate about what I think and write and I hope it is coming across this way and that you will agree with my conclusions.

Lastly before I go too far, I will be covering some very off-the-wall subjects further into this chapter, and as you have read, I have also been covering religion. I just wish to be clear that I am in no way trying to offend or discredit anyone or anything - just purely writing about facts and discoveries that have been made, trying to understand biblical writings and putting out alternative thoughts and ideas on what was witnessed and written about through history that may have been recorded in a way that was incorrect, due to nothing more than a lack of understanding at that time.

Part E: Made in His Image

Let's look at what we do know about mankind's history based on scientifically proven facts.

Homo sapiens were not alone. Long ago, there was a lot more human diversity. Homo sapiens lived alongside a recognised estimate of eight now-extinct species of humans, this around 300,000 years ago.

As recent as 15,000 years ago, we were sharing caves with another human species called Denisovans, a species of humans that were first identified as recently as 2010. Further to this, and through fossilized remains, indications that an even higher number of early human species once populated the Earth before our species came along.

Now we have only one species, we are the only ones left, which historically is very weird.

My first thoughts dwell back to the origins of mankind I covered in previous paragraphs. These scientific discoveries have for me proven that a benevolent God did not create mankind as outlined in the Old Testament, and in whatever religion, if we were made in His image, then where did all the diversity and completely different, at DNA level, human life forms come from? This *made in His image* crops up so much in historical texts and I am trying to unravel why.

Even now with just Homo sapiens remaining, there is still huge diversity. There is a huge difference in appearance between a man of 6.8 ft in height, muscular, with dark full body hair and beard and that of a blonde-haired, petite woman of 5.2ft in height with large breasts. There is a huge difference in appearance between a man of Chinese origins and that of a man of European origins. There is huge difference between a white man and a black man. These differences are only outward appearances of course as we are all the same inside, but the historical references are made referring to image, not what is inside. So, what exactly was this image?

I will share my thoughts on the *made in his image* reference later in this chapter.

My second thoughts are, and based on these many diverse and different humanoid species we now know for sure once occupied the Earth, why is everybody so hung up about the idea that we are not alone in the universe?

Even now this planet is covered in literally millions of different life forms, from microscopic bacteria in the air, to huge elephants on the land, to the enormous whales of the sea. We now know that there was a time very recently in cosmic timescales, akin to about a second ago in the lifetime of the planet, that we shared this world with many other human life forms, all different, just on this tiny planet, so what with the hard-to-imagine number of planets there are in the vastness of space, why do we insist on thinking we are all there is? It only takes one other planet to be in what scientists consider the

Goldilocks Zone, to be able to sustain life as we would know it, to say nothing of life forms we as yet do not understand, so why do we consider ourselves unique and all there is?

When we are talking about races that are different in appearance and image, we must also consider stature, and as such consider another conspiracy theory that is possibly fact, and that is regarding giant beings having lived alongside us in the past. This, because of all the historical texts mentioning giants, wall drawings from all continents and civilisations from history of giants, and the large stature of giant skeletons unearthed over the last couple of centuries. For me, it just adds another blot on what we truly know of our history. Again and on the other end of the scale with regard to stature, the word pygmy describes the rainforest hunter-gatherer populations from around the globe that still exist today that share an average height below 5 feet tall. It is my understanding that the current global average for humans is 5.4 ft for women and 5.9 ft for men, a big difference you will agree between the average human and the pygmy race that again questions *made in His image*. If we have a race of tiny people still existing today, then why not a race of tall people in what was our recent history?

For whatever reason people are generally more comfortable with the understanding that we are all there is just because that is all there is right now.

So, let's acknowledge at least one truth. Very recently in the history of not only Homo sapiens but the Earth itself, the inhabitants of this planet would have resembled scenes from movies like "The Lord of the Rings", with all different types of humanoid life forms of varying appearances and sizes.

In spite of all I have just written people choose to ignore facts to feel comfortable in their existence and not acknowledge the truth as scientifically proven.

Anyway in conclusion to the subject of Homo sapiens being *made in God's image*, it is clear to me that through our scientific and archaeological discoveries and our studies of historical texts, that the percentage chance that we shared this planet with many other human species that all varied in appearance and size in our recent history is far higher and nearer to absolute certainty that most would consider and in most cases want to accept.

Part F: Giants

Further to my comment regarding giants, I feel I should say more about some of the overwhelming evidence I mentioned that would allude to the fact that they were/are real, which I know will raise an eyebrow as they have been mentioned so much in fictional literature and seen in movies. As stated previously, giants are mentioned in the bible. One such reference is to the giant Og, who was the Amorite King, who was defeated by the Israelites as they entered the Promised Land. King Og's bed was made of iron and is noted as to be 13.5ft in length and 6ft in width in modern day measurements, so a giant man indeed if requiring a bed this big! Records show he was the remnant of the Rephaim, who were a people known to be fierce fighting giants - Rephaim is a Hebrew word for giant.

Perhaps the most famous giant of all is Goliath of Gath, the philistine giant defeated by the young David in single combat. Many variations have arisen for the true height of Goliath, based on what a cubit and a span is in relationship to modern measurements (a cubit and span being used to measure in Goliath's times). As such, anything from 6.9 ft to 11.5 ft has been documented through the years as Goliath's true height. Recent investigations and studies believe that David was 5.3ft tall, the average height for a male at that time, and Goliath was 8.11ft tall, almost twice the height of David! As a point of reference the tallest human ever recorded, Robert Wadlow, who suffered an affliction that meant he never stopped growing and that he died at the young age of 22, was also 8.11 ft in height. The difference is that Robert would have been in no condition to be encased in heavy armour or attempt combat; he walked with the aid of a walking stick and had leg braces. This for me, if Goliath was indeed real, further proves that Goliath was not just a very tall man but in fact a completely different race altogether, which is not so hard to believe these days given our understanding of all the different humanoid races that have existed. If we can have pygmies shorter than us then why not a race of beings that were taller than humans.

Trust me when I state there have been numerous things found and documented through history, including most recently, that alludes to a giant race having been here on Earth at some stage - or possibly still here now, choosing to stay hidden!? I do not want to go too far down that rabbit hole as this book is not about giants. You can study this subject with ease for yourselves and the evidence is compelling to say the least. A good place to start is the Patagonian giants, the giant fossilised footprint found near the border to Swaziland in Africa, the recent encounter with a giant by military forces in Afghanistan, or the lovelock Cave Giants in Nevada USA.

That said I will just touch on two more facts I am aware of that are very important to further cement this case for giants, a subject which I am only mentioning in this chapter to query why they are not more prominent in our known history when evidence suggests they once were inhabitants of this planet and walked alongside us.

Abraham Lincoln, regarded as the greatest President of America there has ever been, once stated, when looking at Niagara Falls... *The eyes of that species of extinct giants, whose bones fill the mounds of America, have gazed on Niagara, as ours do now*–Abraham Lincoln, 1848.

This is based on the massive amount of large skeletons that have been unearthed in America over the last century or more, skeletons with two rows of teeth, red hair, and huge stature. You can investigate this further for yourselves and I implore you to do so, and check out the uncanny number of times the Smithsonian Museum are mentioned as having taken them away for storage, or as I like to say, they've hidden them away!

Secondly and again please investigate these yourselves, there are an overwhelming number of wall drawings of giants that have been found around the world on every continent and from every civilisation that existed. Even in Egyptology there are drawings of people who are interacting with beings of giant size - please go ahead and Google these things yourself. Ask yourself this, why would any intelligent person like Abraham Lincoln make a public address referencing giants? Or an artist trying to document what he saw, draw people of normal stature alongside a huge giant if they did not exist? Why would you even consider doing this?

Part G: The Garden of Eden

Now, please bear with me on this next bit as it will take some open mindedness to appreciate my ideas and thoughts on this matter. Earlier I made mention of the Garden of Eden and that I would give my thoughts on the matter later in the chapter. My reason for this was that I felt it was something that needed its own section to do my ideas justice.

Now, with all the various historical texts we have gathered through history, examples being the Dead Sea Scrolls, scriptures like the Sutra, the Bible, the Pyramid Texts, the Book of Enoch to name a few, there is a glaringly obvious similarity with one fact, and that is that a deity of some form is always mentioned. In my mind, these ancient writings must be attached to some form of truth. The old saying is "there is no smoke without a fire". I just feel that due to a major lack of understanding at that time in history regarding our planet, the universe, the complete lack of anything resembling scientific studies, that things that were witnessed and written about were misinterpreted as these would have been seen through eyes that resided in primitive cultures. As such, how could the Garden of Eden have been possibly misunderstood?

I have always felt that we have been visited all through history by off-world, advanced beings, and not all from the same origin and not all being the same species, likewise with various degrees of advanced technology with some being more advanced than others, some simply from other planets in our galaxy, some more advanced and from other galaxies in our universe, and some far more abstract and from completely different dimensions and vibration levels. All with different agendas on their interactions with the populace of planet Earth at the time of their visitations, all of which I will provide compelling evidence for later in this book.

I think this as there is so much evidence to support this idea, far too much to be simply ignored. In fact, there is way more tangible fact-based evidence to support this idea than any of the various religions there are around the world could ever provide to support their ideas, yet still we still flock to pay homage to Gods and deities that thus far we have absolutely zero evidence to support their existence.

Please remember that global governments are in the process of providing disclosure on the existence of advanced off-world beings, have acknowledged that vast sums of money have been spent investigating this phenomena over many decades and have admitted to the existence of UAPs. This disclosure, along with the millions of eyewitness reports made around the world every year, filmed footage etc, kind of makes it hard for anyone to make a compelling argument against my train of thought.

So, how did the Garden of Eden get so misinterpreted and declared as God's action, rather than the greater chance that it was something to do with advanced off-world beings?

Let's look at the serpent that could speak with Adam and Eve. Chinese history is bolstered up on the fact that dragons and serpent-men once existed, the Dragon King for example is revered in China, and serpent-men are also revered as deities, such as Fuxi and Nuwa, described as snakes with human heads. Now, of all the many alien abduction cases and witness reports made around the globe every year, the most common description for their appearance seems to be that of either the small greys, the tall human-looking Nordic race or the reptilian race. With this in mind, could it be that the serpent in the Garden of Eden that spoke to Adam and Eve was just an off-world reptilian being?

I appreciate that the subject of reptilian beings has descended into a laughable joke these days, but when you look at the overwhelming references to this race in places like the Chinese culture, likewise Ireland and St Patrick, Norse mythology, Greek history, and in eastern religions of Hinduism, Buddhism and Jainism where a semi-divine race known as the Naga took a half-human, half-cobra form, it makes you realise there could be so much more to this subject than mere jokes. As I stated earlier, there is no smoke without a fire, and all through history and in nearly every culture on the planet, there is reference made to serpent-men.

In conclusion, could these serpent-men just be an off-world race that we perceived as deities and that one of them spoke to Adam and Eve? Could it be that Adam and Eve were not the first humans here, that this planet was already covered in human life forms, and that Adam and Eve were simply another off-world race like the Nordics, and that we witnessed them simply talking and communicating with the reptilian race? The apple of all knowledge could have been something akin to a laptop, full of information, that Adam and Eve tried to take for themselves. Maybe Adam and Eve were simply asked to leave the reptilian ship upon trying to acquire reptilian technology, not as we tell it as banished from the Garden of Eden.

I am rambling now, but you can see how easy it is, through lack of understanding, to perceive something totally wrong, and I am not saying my interpretation is correct, just chucking one possible answer out there. Even though my ideas could be seen as just plain silly, is it any more ridiculous than a Garden of Eden, an apple containing God-like knowledge, and talking snakes?

Having read many cases of reported human abduction in recent times by off-world beings, there seems to be a similarity in the experience, regardless of what race carried out the abduction, and that is the medical experimentations that take place, the kind that human scientists carry out every day all over the world on lower life forms like animals and germs etc. My gut feeling is that we are being assisted in our evolution, and that this is a benevolent activity. I feel this as over the last 100,000 years, all other humanoid races have been eradicated and Homo sapiens have not only become the last surviving race, but we have also gone from hunter-gatherers to a multi-cultural, global travelling and communicating civilisation, even beginning to make very humble tiny steps in understanding the universe and our place within it.

Would it be so hard to imagine that alien races far in advance from our current understanding of technology and the universe, this as a result of nothing more than that they have been around longer than us, literally go around terraforming planets and seeding life forms on those planets, and after many millennia come back to check progression, eradicate failed experiments (the dinosaurs for example), and in the case of planet Earth, eventually chose one race out of the many humanoid species that were spawned to become the dominant race - to continue with the evolution of that species by continuing the genetic improvements until that race could take its place within the cosmic family.

What if this is something these advanced aliens knew through thousands of similar processes they had undertaken on other worlds over thousands of years, and was something that, if left to natural selection and natural evolution, could take hundreds of millennia to achieve and they wanted to give us a boost along as they had done on thousands of other planets with other species?

What if, in our primitive understanding of all things at this time, we attributed these alien beings to being God, and their ship as being the Garden of Eden, which is described as paradise in the Bible? Well, regarding the surrounding habitat I would imagine the interior of an advanced spaceship to very much resemble paradise to primitive humans. I mildly and humbly suggest my own current apartment would seem like paradise when you consider living conditions back then. What if the tree of life was the only way for primitive minds to understand what could have been incubators for humans to grow at a rapid rate when genetically modified? Would that make more sense given our understanding of incubators and genetic modification nowadays than some tree of life?

I know this all sounds far-fetched, but if we travelled back in time and spoke to primitive humans and told them we would one day fly non-stop from England to Australia on the other side of the world in just a day, it would seem very far-fetched, especially given that a good nineteenth century, (not that long ago) clipper, a fast sailing ship, took four months to complete that same journey. Do you think they would believe you? We could not even fly then, but we've come a long way in advancing our technology since the onset of our technological boom. I tend to believe in an advanced race with advanced technology than anything offered up previously as an answer through religion. The facts are leaning more towards my conclusions the more we discover about our true history. Maybe the Kardashev scale later in this chapter will give greater understanding of why I cling to these ideas.

Part H: Ancient structures from around the world

When people think of structures from our history that defy our understanding of how they were built, most automatically think of the Egyptian pyramids, and rightly so. They are an enigma.

Having worked in construction most of my life, and in understanding we are sometimes restricted in our building methods by modern health and safety regulations that did not exist prior to 1974, I still find the construction of the pyramids a huge mystery, given that there are an estimated 2.3 million stones used in the construction of the great pyramid of Giza, each weighing somewhere between 2.5 and 15 tons. Number one, where did the stones come from? What quarry provided these stones, and how the hell were they put together the way they are? The great pyramid is over 480ft tall - that exceeds a 35-floor tower block in height!

Even more of a mystery is the fact that experts believe through geological studies that the great pyramid of Giza was constructed over a period of 2 decades. That means that over 300 stones were quarried, shaped, moved from the quarry to the construction area, put into place with pin-point accuracy every day consistently for 20 years, all just using human labour. I can guarantee we could not accomplish that feat today with all our technology and equipment, to say nothing of the two pyramids that sit beside the great pyramid of Giza.

Another discovery that tweaks my curiosity regarding our history are the H blocks of Puma Punku.

High in the Andes Mountains in Bolivia, at an altitude of over twelve thousand feet, is the ancient megalithic site of Puma Punku, and it is unlike any other structural

JUST A THOUGHT

phenomenon around the planet eg. pyramids. Puma Punku stands alone and is unlike anything else found on Earth. Scattered over the land are great stone blocks believed to be 2000 years old, each with cuts made to them with astonishing precision and weighing over 100 tons a piece. After decades of study, their origins still remain a mystery as nobody knows how they got there, who made them, and what they were used for. There is no historical background to them at all.

In their present state they are scattered as if blown away by an explosion, like scattered Lego bricks. What force could have produced a result like this when no type of explosive would have existed at the time this dismantlement would have occurred? Clearly at some time way back in history they were all linked together and most probably formed some kind of structure. There are some that still remain connected, and the extraordinary point of that is they connect with such accuracy and precision that you cannot fit a razor blade or human hair between the gaps - constructed to absolute perfection!

The right angles on the stones are perfect, the cuts as smooth as glass to the point you can still cut your finger if you run it across them, and along with perfect drill holes right through them, experts are led to agree that to produce such accuracy today we would have to machine-cut such holes and angles, possibly using laser technology, which as we know did not get invented until the 1960s. No hand tool could produce such polished perfect drill holes or angles, to say nothing of the gaps joining the blocks being so perfect you cannot fit a razor blade between them. Archaeologist Arthur Posnansky, a researcher who spent many years studying Puma Punku, concluded by using the position of the site in relation to the stars, something all ancient monoliths seem to have in common, that Puma Punku is in fact far older than the original 2000 years old that is was thought to be, and it was in fact built in 13,000BC, which if true, further mystifies the H Blocks. Nobody knows for sure the exact age of the site, but for certain the site is one of the greatest mysteries on Earth.

The blocks were quarried from a site over 60 miles away. Due to the altitude, there are no trees growing, and never have, so the usual method put forward by experts for moving blocks of this size in ancient times is squashed immediately. There were no trees to roll them on, so how did they move stone blocks weighing 100 tons over 60 miles, up mountains and hills? Pretty near impossible even today as this next piece will demonstrate.

At the riverside stone quarry in California, they were tasked with moving a rock weighing 340 tons to be used in Los Angeles for an exhibition of art; it was to be called Levitated Mass. This journey was approximately 100 miles over flat land, and took a crane worth £1.4 million, huge hydraulic lifts, a rig with 44 axles, two trucks, one pushing and one pulling with over 2400 horsepower, and all set on a bed of 208 tyres, to achieve movement at 5mph, and one heck of a lot of planning and organising. Going back to the H blocks, although weighing less than Levitated Mass, coming in at 100 tons, if they were built by the hand of man, unlikely given lasers did not exist, then they were moved by hand up maintains over a distance of 60 miles, yet the logistics even today regarding moving huge stones would suggest that this was impossible.

Further to the H blocks and another mystery in history are the huge stones in the Temple of Jupiter in Baalbek, each block in the walls weighs 300 tons, similar to Levitated Mass. In this case though, there are many blocks weighing

JUST A THOUGHT

this much. However, there are 3 blocks that weigh over 800 tons a piece. This trio of blocks has been nicknamed the Trilithon. They are the largest stones to ever be cut anywhere in the world, each being 88 x 48 metres in size, that's 288 ft x 157 ft. Simply put, they are enormous! Levitated Mass weighs 340 tons, arguably the biggest stone movement of the modern era, less than half the weight of the Trilithon blocks. Experts in this field, especially those involved with the movement of Levitated Mass, including the owner of the Riverside quarry, have absolutely no idea how they were moved or even more baffling, how they were put in place with such accuracy. 800 ton rocks are near impossible to move even with today's technology.

With the best equipment money can buy today, with cutting edge technology, we cannot construct some of the ancient huge monoliths that still stand around the globe today, so my thoughts on history regarding these huge construction projects are....who built them and how?

Another construction mystery are the ancient structures build into solid rock that can be found all over the world. The Kailasa Temple in India, carved vertically from the top down in solid rock is still a huge mystery even today. Archaeologists suggest this method of construction in solid rock should have taken approximately 100 years, yet in reality it was completed in just 18 years. Engineers of today find it impossible to achieve this feat using modern technology in such a short period of time.

Another mystery are all the cities found under the seas. The mystery is not why they are submerged, that can be explained by the last Ice Age and the melting of a frozen world that caused sea levels to rise dramatically. No, the mystery lies in the size and complexity of them. Some of these cities are on par with the Roman cities that were to follow thousands of years later. Again, who built them and who had that kind of knowledge when really the history books say we were hunter-gatherers at that time?

Anyway, there is a lot of source material on this subject in books and on-line, but this brief section just further cements my idea that we know nothing of our past really, with so many mysteries still outstanding in just the construction world alone. Why were they built? For what purpose? How were they built? And by whom? Just adds to my issues with our history.

Part I: My Conclusions

Personally I think there is so much we do not know of our past. I feel that scientific discovers are changing our perception of our history with every new discovery happening almost daily at present.

We do know that we have inhabited this planet longer than we thought. We know that we shared this planet with other humanoid life forms of various sizes and appearances. We know there have been many extinction level events on this planet. We know that there are structures around this planet that we would struggle to build or still not be able to replicate even today, so how did they get there? I can state with reasonable confidence that we have been visited by off-world beings all through our history. We know that the religious explanations for our past are becoming more than questionable and not holding up against today's scientifically proven facts.

So, I guess in conclusion, I have no more idea about our true history and past than I did when I started this chapter, and that my issues with history still remain. All I can hope is that you, the reader, will look into this subject further for yourselves, and if I have tweaked somebody's curiosity after reading this chapter, which is what I am trying to achieve, then maybe someone one day will come up with the answers and facts about the true history of our existence.

Part J: The Kardashev Scale

As a last thought, within some of the above, I have kind of beaten up religion in places, and as said, that is not my intention. It is just that religion has laid claim to being the architects and builders of our beginnings, and that as we develop as a species and our understanding of everything grows, religion has consequently left itself wide open to questioning the validity of what they have claimed about our beginnings and God. I am just writing things as I see them and referencing facts that question religious explanations. I have no issue with anyone who follows any religion or faith. I truly am a live-and-let-live guy. As such, I just want to cover one last titbit of information that will possibly bring us full circle back to religion in a positive light, with regards to our origins, albeit with a slight twist...

Back in 1964, Soviet Astronomer Nikolai Kardashev came up with a hypothetical scale for measuring a civilization's level of technology based on the amount of energy it is able to use and control. This scale had three base levels or types. I say *had*, because the scale chart has been added to since its conception which I will cover later. Briefly, without all the finer details, this is what the three scales equated too.

Type 1: is given to a species that has been able to harness all the energy that is available from a neighbouring star (using solar panels etc.) and be able to gather that power and store that power for the growing demands of a growing population. This civilisation would also be able to control all natural forces on their planet by having control over the weather, volcanos, earthquakes etc.. They would be able to traverse their own solar system with ease.

Type 2: is given to a species that can harness the power of their entire star and store it on a planet for use by that civilisation (not merely transforming starlight into energy, but controlling the actual star). At this stage in the evolution of this species, no known threat could wipe out a type 2 civilisation. If a moon-sized object entered their solar system on a collision course with their planet, they would have harnessed and be able to control power that could literally vaporize it out of existence. This would also be an inter-stellar race capable of travelling to other nearby star systems.

Type 3: would be a species that has become advanced enough to be able to traverse their entire galaxy (our Milky Way for example), with knowledge of everything there is to know with regard to energy and having the ability to harness all the power from any chosen star in their galaxy. They would be capable of colonizing any planet within their own galaxy. The galaxy would literally become their playground.

The Kardashev scale is widely recognised as the principle scale for a species' technological progression and current state, and used globally by many different administrations as a form of reference. When we consider the human race and our own technological placement in this scale, it really drives home that we are just starting our technological journey when it comes to technology and our advancement as a species. Since Kardashev first came up with his scale, 4 more types have been added as we furthered our understanding of our universe and probable existence of a multiverse. One of the new types is Type 0, which we fall into at present: a civilization that harnesses its energy requirements from its home planet with minimal harnessing from its neighbouring star, basically still reliant on fossil fuels. Currently we are at about 0.73 on the Kardashev scale and it is presumed by scientists that we will reach Type 1 in about 100 years, give or take a few decades depending on how fast technology advances and how diligently we procreate.

Type 4: is a species capable of harnessing all the power of the known universe, capable of being able to travel around the entire universe at will, travelling and visiting every galaxy in their universe, capable of consuming the power of several - possibly all- galaxies if required. This civilisation would be capable of manipulating space time and playing with entropy, thus reaching immortality on a grand scale, essentially an indestructible and a highly utopian civilisation.

Type 5: would be a multiverse culture, capable of harnessing the power of multiple universes across multiple dimensions. A civilisation that had outgrown its own universe and had to span out over other parallel universes, a civilisation capable of manipulating the very structure of reality.

Type 6: is even more abstract than Type 5. This species would exist outside of time and space as we know it, and would be capable of not just controlling and harnessing power from multiple universes and dimensions, but literally having capabilities to allow them to create universes within the multiverse, and destroying them just as easily. Akin to a deity or god perhaps and would be beyond anything we could comprehend at this time.

These are the current types of civilisations represented in the Kardashev scale as of 2023.

The description of Type 6 could allude to a deity or god who could in fact have created everything as outlined in religious writings, describing a civilisation that had passed through unfathomable amounts of time, and gone through their own evolutionary journey. Any race that progresses up the scale must evolve from their previous type, amounting vast knowledge and eventually leading to knowledge about literally everything there is. In other words, a Type 6 species would have once been a Type 4 and Type 5, both of which were capable of manipulating space time. Type 5 also being able to manipulate the very structure of reality, both Types being able to play with time allowing them to occupy reality whenever they wanted to which allowed them extended time to evolve into a Type 6 species living outside of time and space as we know it. A species that, if we ever came across them, we would, even in today's modern world, rightly refer to them as God, with us not being able to remotely comprehend or understand a species so advanced. Even now with our humble understanding of how the universe works, they would be so in advance of us we would literally consider them God or god-like.

Through their evolutionary journey, this species would have acquired so much information that their existence and knowledge of everything would be beyond anything any other civilisation could understand, outside of a Type 4 or Type 5 species. Even a Type 3 species that has the knowledge to allow them to traverse and harness the power of an entire galaxy, would not be able to grasp the concept of a Type 6 species. If a Type 6 species tried to explain to anyone of these lower types of civilisations, the details of their knowledge and existence, it would be akin to us trying to explain to a single celled organism, in every detail, how a helicopter is constructed, and how it works! To give you a further idea of how far in advance they would be to all other life that exists, the Star trek universe that I am sure we are all familiar with and would seem to us to be far in advance of where we are right now would only be considered a Type 2 culture, a species capable of travelling to nearby star systems but not yet capable of traversing their entire galaxy like a Type 3.

It would simply be impossible to educate any species from Type 0 through to Type 3 of the basics of their existence, so why would they even try to do this? They would have knowledge so far in advance of what we understand, we could not even begin to take it on board. Not just a few hundred generations ahead of us but literally billions and billions of years ahead. When you consider what we have achieved in the last 80 years in our technological boom, imagine carrying on at that rate of advancement for billions and billions of years. It's hard to imagine how advanced we would become.

Which brings us nicely back to religion and that there could be some truth to a benevolent god, that God as we understand it is possibly just a species that we perceive as God, creating everything, a species that has evolved over many billions of years to a point that they are now god-like compared to all other Types on the Kardashev scale regardless of what number they occupy - Ultimate knowledge of everything there is, able to control, create, and destroy everything there is, residing outside of time and space as we know it - sounds very much like God to me.

If the scale is right and there are civilisations that occupy all Types on the scale, and we have interacted with a Type 6 in our history, it could explain our writings about a god or historical deities in other cultures around the world.

One last curiosity, through the many billions of years to come as we hopefully rise up and through the Kardashev scale, will we be adding more Types to this scale as we rise up through it and discover further races far in advance of and so abstract that even a Type 6 species pales before them?

Part K: An Afterthought

Has anyone else noticed that we humans do not seem fit for purpose on this planet? By that I mean we are pretty much the only thing on this planet, living, that does not seem to fit the environment we live in. Briefly, if it's cold we freeze, we only need a slight bit of sun for us to burn or require us to wear sunglasses to protect our eyes, if we go under the sea we drown. There does not seem to be one single environment on this planet that suits the human race, yet everything else on this planet fits into its environment without issues.

A camel can survive the blistering heat of a desert, a fish can survive the crushing ocean depths and 'breathe' under water, a polar bear can survive in the freezing conditions of the Arctic. Where is our environment if we did in fact evolve on this planet? Surely if we evolved here, we would be fit for this planet's environment.

I will let you mull that one over.

Chapter 4: Are we Alone?

Are we alone? That has got to be one of the single most asked questions of the last 100 years, and it refers to us as a species, are we alone in this universe?

Due to the re-branding of UFOs into UAPs and for clarity in this chapter, I will refer to older cases as UFOs as that was the correct terminology for the time, and more recent as UAPs, or in some cases both.

Personally, it is my belief and I am 100% sure about this, that this universe, multiverse, and all the different vibration levels to which life exists are absolutely teeming with life.

Before I start outlining some of the finer details of what we do know, let me hit you with some big and well known facts regarding this subject.

Thousands of UFO/UAP sightings are reported every year around the globe, not including the thousands that go unreported, from all walks of life and from every continent. In 1947 the U.S military started to investigate this phenomena, but supposedly suspended the programme in 1969 after looking at 12,600 individual cases. The investigators were left with over 700 incidents that after looking at every possible explanation, they could not explain these 700 incidents at all, that is a frighteningly high number.

Every civilisation on every continent all through history have references to flying vehicles and advanced beings with advanced technology, including many religions - what can't speak can't lie!

Many famous people and some of great credibility have reported seeing a UFO/UAP and had contact with alien lifeforms, to name but a few:

Alexander the great observed "gleaming silver shields overhead" in 329BC.

John Lennon reported seeing a UFO outside of his bedroom window.

Muhammad Ali once saw a UFO while on one of his early morning runs, just prior to dawn. In fact, when he talked about this incident, he went on to say it was his 16th sighting and that he regularly looks up to the night sky to see them.

A few American presidents have stated they have seen UFOs, the most famous being Jimmy Carter, who made no attempt to disguise this fact, and went into office vowing to disclose the issue. He never achieved full disclosure but more classified documents relating to the phenomena were released in his presidency than any other president in history.

Let us briefly look at those people:

The most accomplished military leader in all history.

One fourth membership of the biggest band in history.

Arguably the greatest boxer of all time, voted the greatest sportsman of all time, and one of the most famous human beings ever to live.

And off course the president of America.

I find it hard to believe that any of those guys needed additional attention or reason to extend their personal fame if these statements were made for attention seeking purposes, or to discredit what they have publicly and without shame stated about their own experiences with UFOs.

In 2017, the New York Times published a front-page story that the pentagon was still conducting a secret program to investigate pilot encounters with airborne objects that displayed un-earthly speed and manoeuvrability. Shortly after this publication the TIC-TAC footage was released into the public domain and full disclosure about the fact that UAPs are in fact real was released by the U.S Government. This further cemented public perception that the U.S Government, and other governments and administrations around the world know more than they let on.

Recent studies carried out in America showed that 65% of the U.S population believe in alien life beyond the Earth. That is a staggering number and will undoubtedly include all manner of people, some of whom will be doctors and nurses, teachers, lawyers, business people, police officers, military staff, politicians, building tradesmen, sports stars, architects and designers. I guess what I am saying is that that number will include some very professional people who are held in high regard in society, people not prone to flights of fancy who could possibly jeopardise their careers by openly stating what they believe in, and I would also guess that if these sort of studies were undertaken in other countries, they would find much the same results.

Surely, we can no longer just approach this subject with a bull-headed, not-prepared-to-discuss-it attitude and that it's all rubbish and remains a taboo subject. It has gone beyond personal interpretation now. The facts indicate that there is so much going on regarding this subject that we must start taking it seriously. Unknown craft that are hundreds, maybe thousands of generations ahead of us, flying into our air space un-challenged, thousands of reports with some startling evidence to bolster up the claim of human abduction, and none of us seem concerned or take it seriously.

If just one, un-challenged craft of advanced nature not from this world is flying in our airspace, or just one human abduction has taken place, that should be enough to precipitate an all-out global multi-nation investigation, yet these things are reported by the thousands every year, and as pointed out, by some creditable people, statistics suggest that the phenomena is believed by approximately 65% of the population, and nothing happens? I would argue it is up there with global warming as the number one subject that requires discussion and action sooner rather than later, and if our world leaders do in fact know what is going on and are withholding that information, is it not about time that it was all disclosed to the population of the world? We have a right to know the truth.

Anyway, that's just a few bullet points, let's look at things in more detail. I think the best place for me to start is way back in history, and of course I am not going to write about or catalogue every single piece of potential evidence to support the idea, just some key things that cannot be ignored.

As you will be aware, I covered some of the structures around the world that we could not replicate even with today's technology in the last chapter, and I do not want to go over the same ground.

Other than to say, seriously, I implore you to look at some of these structures, Google them. There are literally so many examples around the world, and if we cannot build them today at the rate that geologists state they were built, through decades of investigation, then who did? If we cannot form and move blocks at the rate that we believe they were moved at for construction purposes back in our past, to say nothing of the blocks we would struggle to move even now as they are so big and heavy, then who did? If blocks were clearly cut to such precision that only a laser could achieve this, or contain drill holes so perfect that only a diamond tipped drill could achieve this, thousands of years ago when we were still using crude hammers and chisels, then who did this work and how?

I am going to be using a court room process/procedure to make my points through this chapter, e.g. present the evidence as we understand it and go with the obvious conclusions. Given what I've just written, the overwhelming evidence would indicate that these structures were not built by the hand of man. If anyone can offer up another suggestion as to how such incredible structures were built all that long ago, then I would love to hear it. But given that after centuries of study, nobody as yet has offered up a realistic explanation, I would say case closed. How much longer are we going to kid ourselves or turn a blind eye to the glaringly bloody obvious, somebody else built them and with advanced technology that we have only just recently invented for ourselves through the period of our technological boom, and in some cases in advance of what we have even today.

Conclusion: Off-world, advanced, by even our modern technological standards, alien beings were responsible.

History is literally littered with references to encounters with other worldly beings, sightings of flying ships, structural anomalies, crop circles etc, yet our history is replete with miss-information. Oh, and contrary to common belief, crop circles are not a recent phenomenon. They are recorded as far back as the 17th century, reported by a farmer who witnessed a dazzling light over his oat field at night, only to find in the morning that perfectly round circles had been formed in that field.

One of the most compelling pieces of evidence from our past that refers to advanced races interacting with not only the human species on this planet at that time, but also interaction between different alien races, inclusive of a war fought out on this planet between advanced aliens, is to be found in writings in the *Mahābhārata*. This ancient Hindu text contains writings that you would think came straight out of the pages of a modern-day conflict, not written 2500 years ago! These writings contain what would seem to be references to missiles, atomic weapons, huge laser weapons that could melt entire cities, all manner of weaponry. The question is, why have some of the earliest accounts of warfare have within them references to sophisticated weaponry that we humans would not develop for thousands of years? Why is the *Mahābhārata* littered with references to weapons that are strikingly similar to what we have today?

The god Vishnu had a weapon that, once launched, was equipped to find its target. Sounds very similar to a guided missile. There is reference to weapons that were heat-seeking, again similar to our current heat-seeking missiles. In fact, in the *Mahābhārata*, approximately 46 different types of weapons are described, each having a different function. There are descriptions of flying vehicles called *Vimānas* that can disappear - would this be some form of cloaking device, or stealth technology like we have now, or simply just disappearing through a portal, giving the illusion of disappearance? Weapons that could put people to sleep, would these be nerve weapons? We can find writings that refer to the distance of a light year, or the length of an atom, only a society in possession of nuclear energy would have such words. I think the scariest writings are in reference to what sounds like nuclear or atomic weapons.

Consider these verses from the ancient *Mahābhārata*:

A single projectile charged with all the power of the universe. An incandescent column of smoke and flame as bright as a thousand suns rose in all its splendour. The cloud of smoke rising after its first explosion formed into expanding round circles like the opening of giant parasols.

It was an unknown weapon, an iron thunderbolt, a gigantic messenger of death, which reduced to ashes the entire race of the Vrishnis. The corpses were so burned as to be unrecognisable. The hair and nails fell out, pottery broke without apparent cause, and the birds turned white. After a few hours all foodstuffs were infected. To escape this fire, the soldiers threw themselves in streams to wash themselves and their equipment.

A thick gloom swiftly settled upon the hosts. Fierce wind began to blow upward, showering dust and gravel. Birds croaked madly, the very elements seemed disturbed.

The Earth shook, scorched by the terrible violent heat of this weapon. Elephants burst into flame and ran to and fro in a frenzy over a vast area, other animals crumpled to the ground and died.

The destruction was caused by...a single projectile charged with all the power of the universe.

Does this not sound exactly like what a nuclear/ atomic bomb would be described as, like when used? Exactly as the Hiroshima and Nagasaki bombings have been described?

This was all written 2500 years ago, long before we even had the concept of such weapons or even considered such technology, long before the Wright brothers took their first flight. Yet the Mahābhārata writings refer to a great war that was fought, not only on land but in the sky, by flying vehicles, and use of highly advanced weapons and weapons of mass destruction. Even the great science fiction writer, Jules Verne, could not imagine such extraordinary weapons and vehicles, and his work was written just over a century ago, not thousands of years ago. Based on the 'what can't speak can't lie' basis of these writings, a major part of the Hindu Religion, I would say the evidence is overwhelming that a war was fought out between two advanced species of aliens on Earth in the past.

Conclusion: The overwhelming evidence provided in the writings of the Mahābhārata would lean heavily towards the fact that a war took place in Earth's ancient past, and the weapons used in this war described and written about over 2500 years ago were used by advanced, off-world alien beings, weapons and technology we are only now beginning to replicate for ourselves.

Clues to other worldly beings having interaction with Earth and its population, are evidenced carvings or formations into the ground. Whilst these are not entirely fact-based texts as like the Mahābhārata, they are no less interesting and give food for thought. For example, and very much like the structures mentioned previously, incredibly sophisticated construction techniques must have been used to carve out on the ground the Moray (Inca ruins), the Nazca Lines in Peru, or to a lesser extent with their longevity in lasting, but still on the ground, are crop circles. My two most favourite ground-based artifacts are the Nazca Lines and crop circles, and as such I will cover what I know of them here.

The Nazca Lines are a group of geoglyphs made in the soil that cover an area of 1,000 sq kilometres. There are about 300 different figures, including animals and plants, composed of over 10,000 lines, some measuring 30 metres wide, and other than a partial very limited view from a localised hilltop, they can only be seen and appreciated in full from the air. Constructed many hundreds of years prior to man's first flight, who were they made for and why, if they can only be seen from above? And who spent what must have been a huge amount of time carving these lines into the ground that nobody could see or interpret unless they were in the air? Were they made by some primitive race for what would have been seen as a sky god in its flying vehicle? What else would they be for?

Crop circles, as said previously, are not a recent phenomenon. The subject of crop circles has always fascinated me. When they first started to appear around the globe they were few in numbers and rather crude in their appearance, and could with relative ease, allow the sceptics to make a compelling case that they were something manmade. But as time has gone on, over ten thousand have now been seen on every continent and in over fifty different countries on the globe. Reports have been made by land owners of their appearance literally cropping up overnight, or they have formed in a matter of hours, and most amazingly, some are even reported as being formed instantly.

With no sign of footprints in the crops to indicate the approach of a human to make their formation, they have become more and more intricate in their design. Some have mathematical equations in them, or binary codes, and the crops in the fields where they form have literally had their structural integrity altered by means of being genetically modified. This is a process and technology we cannot replicate yet. This left me to ponder, are they really man-made?

Although not entirely convincing evidence, there is even suggestion that crop circles are mentioned in the Dead Sea Scrolls - they make reference to a light that leaves a mark on the ground. Could this be an even earlier reference to crop circles dating back thousands of years?

More recently in our history, things have moved on, like our technological boom, at a staggering pace, and since the 1960s, crop circles have made massive global headlines, with a liberal spreading of circles showing up around the globe through the 1970s and 1980s. This pattern continued until the 1990s when boom!! It all takes off and crop circles are at the forefront of global reporting.

Showing up more and more around the world, and in far more detail in their design, with meteorologists and scientists out in the field all over the world trying to come up with the answers as to what they are, some concluding that they are not of this world, some concluding that they must be a hoax, but for sure it was a time that aroused just about everybody's curiosity as to what they were and what they meant.

This interest and study continued up until 9th September 1991, when two retired men from the English countryside, Doug Bower and Dave Chorley, laid claim to having formed them all. By now the global interest in crop circles had reached fever pitch, and to be fair to Doug and Dave, it was never a better time to gain global recognition, if in fact that was their agenda. They claimed to have formed them using a board and a rope and did in fact prove they could form a reasonable circle of minimal design using this method. Their claim held up until further study and in-depth analysis was made by scientists regarding their methods. The system they claim to have used was deemed not to be able to replicate some of the more detailed designs that had been found around the world, and even the basic circles they had formed were off kilter and far from being a perfect geometric design, like all the other circles found, and could they really have formed all the circles on every continent on the globe?

Their claim lasted for a while but was found to be fraudulent, and some say they were government stooges set up to try and discredited crop circles and bring the whole hysteria to an end. Doug and Dave themselves were eventually discredited as frauds and went on in global media to be called the "men who conned the world". Now, although a small majority have been proven to be hoaxes and manmade, startling evidence has been found and for the most part kept quiet, that proves the majority of crop circles are made by intelligence far ahead of anything mankind has at present.

This evidence is acquired from not studying them from above and looking for the aforementioned geometric anomalies, but from studying them at ground level.

When formations have been made by man, and this has been conducted in scientific experiments, using the only way we can at present form them, using Doug and Dave's rope and plank method, this method leaves physical damage to the crops, snapped stems, kinks in the stems and destroyed flower heads - sure fire signs of a crude and physical demonstration of how to form a crop circle, i.e. manmade. This to say nothing of the footprints left in the crop when approaching the area of work using the rope and plank method. No footprints to gain access are found on the approach to crop circles that remain enigmas.

In nearly all cases of crop circles, from all around the globe, and the part that is not made so public, is that they are not mechanically produced. They are made from a far more refined and technological method that we at present cannot replicate - none of the stems are damaged or snapped, the flower heads are still intact etc. I will not bore you with all the science behind these findings, other than to say please look at this yourselves. We really do not have the advanced technology required to produce these formations for ourselves. Some of the more creative and astounding crop circles that have been formed "overnight" contain binary codes and examples of (mathematical) pi, yet an effort was made to hoodwink us into believing that two men travelled to every continent on the planet and created thousands of these formations overnight using a rope and a plank of wood!

Conclusion: Given the level of technology and the amount of work that has gone into the formations on the ground, from the Nazca Lines of Peru and the crop circles from around the world, again it would hint to off-world advanced beings having an input. At minimum, they are being formed by other means than the hand of man.

So, having now covered just a few things from the past pertaining to this subject, let's bring things forward to our more recent history.

JUST A THOUGHT

During World War Two, many pilots from both the allied forces and the German Luftwaffe all claimed to have seen in the sky, and to have had close encounters with balls of light flying near them at high speed. After the war, it became apparent that all sides had witnessed the same thing and in frighteningly high numbers. They came to be known as Foo Fighters during the war. This phenomena was mostly ignored by all sides during the war but has been well documented since the war ended.

The first real reported sighting of UFOs was made by commercial pilot Kenneth Arnold, who claimed to have seen nine "flying discs" in formation over Washington State in the U.S on June 24th 1947.

Based on the surrounding habitat, which was mountainous, giving him fixed points to gauge speed, he estimated they were travelling at 1,200 mph. Even the first-generation jets of the time could only reach speeds of around 500 mph. The sound barrier itself, 770 mph, was only broken in October of 1947, 4 months after Kenneth witnessed his flying discs. So, nothing man-made would have come even close to speeds of 1,200 mph for a long time after this incident, many years in fact. Kenneth stated that there was no visible propulsion engines and they had no wings, and that they were disc shaped. This report travelled the world very quickly and in media they became known as flying saucers. This is where that term came from.

Conclusion: Given the number of reports made by both sides during the war of interaction with Foo Fighters, can we even begin to argue with the fact that they obviously existed? Based on the fact that Kenneth Arnold was a perfectly sane, well-respected pilot of many years, and up until that point UFO reporting did not exist, so he had no other precedence or examples to go along with (ie. join in with others who had made their statements about seeing flying discs), and that the term flying saucer became the terminology for unidentified objects in our skies for the following decades after this incident, I would say his report and statement is fact-based and sound.

One of the things that strikes me as a concern to say the least and is certainly not an evenly balanced approach to witness accounts when it comes to this subject, is how witnesses are immediately discredited and made to look foolish for reporting sightings, even if one object is identified and seen by multiple people at the same time, including people such as police officers and military staff. Still, every effort is made to dismiss the reports as false or somehow miss-judged and is attributed to being something else as an explanation. The problem is that some of those explanations are more farfetched than the original sighting of a UFO!

Please understand that most reported cases are explained away as being something else and the reasonings are perfectly acceptable and rational. But, if after extensive research has taken place and no other rational explanation is found then we glide into the realms of stupidity with such explanations as a flock of birds in flight, the Northern Lights, cloud formations, shooting stars, even though in some cases the phenomena was witnessed as hovering silently. Even the planets Jupiter, Venus, and Mars have been cited as being an explanation for a UFO sighting in the past. Yet in a court of law, we take witness statements as fact, and as little as two or three witnesses can be enough to send someone to jail for a very long time. Despite this process, a UFO/UAP can be witnessed by literally dozens of people, in some cases in different towns etc, all with corresponding reports and yet they are somehow made out to be fictitious and fabricated.

JUST A THOUGHT

I know of many people who have seen things in the sky that cannot be explained - people I know who are not the type to drift into flights of fancy and certainly not liars. One of the most compelling modern accounts about encounters with not only UFOs but alien life forms themselves, falling into the category of a *"close encounter of the third kind"* within the definition created by famed UFO researcher J.Allen Hynek's scale[1], refers to the events surrounding Travis Walton and his work colleagues.

Travis Walton was an American forestry worker, and on November 5th 1975, while working in the Apache-Sitgreaves National Forest near Snowflake, Arizona, Travis went missing for five days and six hours. While travelling home after a day's work with six work colleagues in their truck, they allegedly encountered a saucer-shaped object hovering over the ground approximately 100 feet away from the road. The saucer was making a high-pitched buzzing sound. Walton was the only person to leave the truck and approach the saucer. It was during his approach that a beam of light coming from the saucer struck him and rendered him unconscious. This frightened the other men who immediately drove away from the scene.

After driving away for a brief time, they collectively decided to return and look for their fallen colleague, but upon arriving back at the scene, the saucer was gone and so was Travis. His colleagues reported the incident to the Police,

---

[1] ***Close encounter of the first kind**: Someone simply observes a UFO but it leaves no evidence.

***Close encounter of the second kind**: A UFO leaves physical traces it was present, such as burns on the ground or broken branches.

***Close encounter of the third kind**: A person makes contact with a UFO or other life form.

and an extensive search took place including scent dogs and helicopters, but there was absolutely no sign or detection of Travis anywhere. His colleagues came under suspicion of foul play and possibly the murder of Travis, as their story sounded so far-fetched, but after they all submitted to and undertook a lie detector test, which are proven to be within 80% to 90% accurate, all his colleagues were found to be telling the truth as they believed it to have happened.

Travis Walton claims to have been abducted by the crew of the craft and found himself in a hospital-like room aboard the craft, where he was subjected to medical experiments and where he encountered what looked like a human being in a helmet, along with other life forms clearly not human, who were 'short, bald creatures'. He claimed to remember nothing else until he found himself walking along a highway five days later with the flying saucer departing above him.

Travis Walton was found to have signs of puncture wounds on his body akin to needle injections, bruising on his body and signs of radiation on his body too.

In all the years that followed on from this incident, neither Travis or his colleagues have changed their story at all and have all been subject to further more sophisticated lie detector tests. In all cases, they passed these tests and have not wavered from their story in the slightest.

Conclusion: Given the extensive search of the surrounding area undertaken by professional trackers with scent dogs and helicopters in support, Travis missing for over five days with no food or water, the medical injuries his body had sustained and the radiation levels, the multiple eye witness reports by his colleagues and the multiple lie detector tests passed by all involved, I would suggest there is real credibility to this incident and the facts point to the possibility that it happened exactly as Travis and his work colleagues have unwaveringly reported for many decades now.

Travis Waltons story was so compelling that it led to a 1993 American Biopic film called *"Fire in the Sky",* a film that was favourably received by critics and the public alike.

As said, I personally know people who have witnessed some profound things regarding this subject. My own mother witnessed UFOs back in the 1960s. What she witnessed in her home town of Hastings was also seen by multiple people as the UFOs travelled down the south coast towards the West Country. Eventually they were drawn by an artist in Cornwall and featured on the 10 o'clock news that night. I know my mother has not lied about what she saw. Are we to believe that all these people from all different places down the entire length of the south coast, who knew nothing of other people witnessing the same thing, whose accounts of what they saw all matched, were all lying or dreaming it up? How could you co-ordinate such a hoax? Even now, that would be impossible.

A good friend of mine not only witnessed multiple UFO/UAPs sightings at once, but it was during the day and he managed to film it. He was so visibly moved by what he had seen he conducted his own experiment with his own filmed footage. He slowed down the footage by 90%. What he found was breath-taking and frankly incredible. The footage showed multiple objects zooming across the screen, also what looked like portals opening and disrupting the formation of the clouds, and one round silver object, the initial sighting that drew his attention, literally split into two parts. This footage can be found on YouTube as my friend uploaded it there, titled "UFO filmed in Westfield near Hastings Sussex".

Something that crossed my mind with this footage was the fact that the objects zooming across the sky cannot be seen at normal speed. They are literally zooming across the TV screen even though the footage is 90% slower than that of normal time. So, is it possible that our skies are filled with these UFO/UAPs all the time and they move at such incredible speeds we just cannot see them?

I have pondered this many times as I look skywards at our aeroplanes flying across the sky. When you consider they look to be moving very slowly when seen from the ground, and are in fact travelling well in excess of 500 mph, you have to ask just what speed are these UFO/UAPs going if they are zooming across the sky at breath-taking speed and we can only see them by slowing down video footage from normal time by 90%? I have looked into this matter and found that some reported UFO/UAP sightings made by all manner of people who had the technology to track them, radar operators, military and commercial pilots, stated that the capabilities of these objects are many generations ahead of anything we possess right now.

The average human will pass out and go unconscious at 9-Gs. G-force relates to the increase in gravity - 1-G is equivalent to normal gravity, 9-Gs is equivalent to 9 times gravity etc. The most up-to-date, fifth generation fighter jet will start to come apart at about 15-Gs. Some UFO/UAPs have been tracked travelling at incredible speeds way beyond what our fasted known jets can reach and then make turns at complete right or left angles all while continuing at these incredible speeds, a feat that would generate G-forces in excess of 1000-Gs.

These objects have also been tracked coming in from space, travelling down through our atmosphere and then entering our seas. That is three completely different environments mastered by one vehicle, for want of something else to call them. Although a case could be made to say we have conquered those environments, I would not say we have mastered them yet, just that we have made vehicles that work in all of those environments. That said, we have not been able to create one vehicle that could work in space, air and sea, without any adaptions and maintaining capabilities beyond anything we currently possess in all three environments. That is an accomplishment many generations ahead of our current understanding of these environments and their working conditions and our technology to produce such a vehicle.

We do not have vehicles that can hover silently, as has been witnessed and reported thousands of times. The nearest we have to that feat is a helicopter, but they are very noisy and most people can identify a helicopter with ease. We do not have vehicles that can go straight up vertically, as has been witnessed and reported thousands of times, from a standing start to being out of sight in the matter of a second or two, with no visible propulsion engine and making no noise. A fighter jet with its after burners on could achieve a vertical climb with some haste, but certainly not in silence and from a standing still hover start.

We simply have nothing that can simulate or match what has been a phenomena witnessed the world over by many people, not merely a few decades ahead of our current technological state, but many, possibly hundreds, maybe thousands of generations ahead of us right now, and yet the whole thing is still shrouded in secrecy and mystery - stories of secret government agencies not only knowing about who and what they are, but working with, interacting with, and forming agreements and allegiances with life forms not from this world or even this dimension.

Conclusion: Given that this phenomena has been documented since our history began, to the point that statistics show 65% of the population believe in them, the structural anomalies the world over, foo fighters during the war, the vast amounts of reported sightings and in some cases interactions and abductions that are reported, the uncanny similarities in the reports made by people who had no idea that other people were witnessing the same thing at the same time hundreds of miles apart, tangible evidence like burns on the ground and increased radiation levels, video footage, wounds to the body after abductions, and radiation burns to the body, crop circles, released footage and admission by agencies that UAPs exist, I conclude that other life-forms exist from other worlds.

Something is definitely happening on our planet, possibly the biggest and well-kept secret in mankind's history, one that could change the outlook of the human race forever, having effects on religions and faiths the world over, finally giving mankind the answers to why we are here and how we got here. If revealed, not only will it discredit every government and agency the world over for keeping this secret from the rest of humanity (no elected official or state funded agency has the right to do that), it could possibly facilitate a complete change in how we operate as humans, bring an end to money and capitalism, end all wars etc, and as such I just wonder how long a lid can be kept on this and when full disclosure will happen. I personally think that this is inevitable.

For sure and I have only just touched on a few things in this chapter that have been referenced or reported through our known history, and unlike in previous chapters where I only sought to challenge the mainstream thinking, here I can say with the upmost confidence, we are not alone in this universe.

Chapter 5: The Blue Marble

In this chapter I am going to try and give you some understanding of how incredibly small and insignificant we are, not only as a race but as a planet (Earth) in the vast universe in which it sits, and if for no other reason than to give some real credibility to what was outlined in the previous chapter. Are we truly alone in an unimaginably enormous universe? As I said, this chapter perhaps better explains how incredibly small and insignificant the human race truly is.

Some of what I write about, like previous chapters, I implore you to pause reading and Google what I am what I am trying to convey. Our existence in this universe is beyond what we would consider microscopic and could almost be considered non-existent.

I have called this chapter The Blue Marble, in reference to a photo taken by the crew of the Apollo 17 space craft on their way to the moon on December 7th, 1972. The photo was taken from a distance of 29,000 kilometres (18,000 miles) and shows our planet as resembling a beautiful blue marble. It is one of the most reproduced photos in all history and has given us some real understanding of how small and precious our home is.

This picture though, was nothing to the photo taken from Voyager 1 on 14th February, 1990, from a distance of 6 billion kilometres (3.7 billion miles). That photo was called

The Blue Dot and was, and remains, the furthest away picture ever taken of planet Earth. The photo was the true awakening of how tiny we are ... planet Earth is literally a dot ... in fact, if not for experts who were aware of where our location should be, you would have struggled to find us! Literally equivalent to a single pixel on the screen, a speck of dust caught in the sunlight. Please appreciate that this photo was taken within the confines of our solar system. Our entire planet, with all its history, absolutely everything mankind had done all through history, all our inventions, all the life forms we share this planet with, and the life forms been and gone (the dinosaurs for example), our vast oceans, high mountain ranges, everything we are and have been, is just a speck of dust on the screen.

Our continuing understanding of our solar system residing within our galaxy which itself resides within our universe, is constantly giving us moments of not only pause, but literally breath-taking awe to what is truly around us and the sheer vastness of everything. If we look at the progression of science fiction to begin with, we can very easily see that our understanding has progressed on a vast scale.

Take Jules Verne for example, who wrote the classic "Around the World in Eighty Days" when the Earth was still considered huge in 1872. To circumnavigate it would be a huge challenge in the time-scale of eighty days with the methods that were available and he used at the time – it still would now - but obviously with today's air travel I am most confident you could circumnavigate the planet in 48 hours, and if you could afford it, travel by first class so the trip would not only be achievable with 78 days to spare but in absolute luxury, and almost effortless! A far cry from the struggles encountered by the fictional character Phileas

Fogg, the main protagonist from the book. Verne also wrote the book, "From the Earth to the Moon" in 1865. I guess at the time of writing, a trip like that was science fiction, only to become science fact a century later.

Thinking of science fiction, early alien invasion stories referenced that the danger would come from the moon. Later however, as our understanding of our solar system grew, that danger changed to a Martian invasion coming from Mars, the most famous telling of this story would be by H.G.Wells, and his "War of the Worlds", where Earth was invaded by Martians from Mars, who possessed advanced technology and weapons. This story was originally set in a rural part of England and later in the story included London, but it has since been adapted many times over and has appeared featuring heavily in America and all over the globe.

More recently and again with our understanding of our surroundings, science fiction has had to look outside the confines of our solar system for alien invaders to Earth. The best example I can think of right now is the 1996 movie "Independence Day", where Earth was attacked by Aliens from outside our solar system who possessed technology far in advanced of our own, city-sized ships and shielded fighter craft etc. So, as you can see, even science fiction has had to adapt to accommodate our understanding of our surroundings, and what can no longer be passed off as fiction has itself evolved to become reality - that change of perception I spoke about in the Preface.

I think to give a greater understanding of what we do know, I will start from my own country and move out to our planet Earth, then our solar system, galaxy and so on and so forth. This will be my approach in this chapter.

Let me be clear though, although we have an almost full understanding of our solar system now, experts state that our understanding of our galaxy and the greater universe is still almost all guess-work, amounting to about 1% fact. So, although we know it is all there and how vast it is, we still have a long way to go before we truly understand its true nature. So much of what we have discovered in the true far reaches of space is still all guess-work at this stage.

I often consider the vastness of everything when I take a long drive. I have a large family scattered all over England, and as such take reasonably long journeys on a regular basis. Examples being to my brother, who lives about 210 miles from me. That journey can take anything up to 4 hours to complete by car, so it can feel a very long way to travel, which I guess it is. That said, when I look at a map of the world and see how tiny England is, and how far my so called long journey actually is, it is no distance at all. As the crow flies, our planet's circumference is 40,075 kilometres (24,901 miles), so my 4 hour 210 mile journey is nothing really, and when you consider Earth's surface area, which is just over 510 million square kilometres (316 million square miles), my journey is definitely nothing at all. Planet Earth is clearly a very big place, or is it?

There is enough space between us and our moon to fit 30 Planet Earths. Our planet shares this solar system with 7 other planets, in distance order moving away from the Sun, you have Mercury, Venus, Mars, Jupiter, Saturn, Uranus and Neptune. Earth is the eighth planet and fits between Venus and Mars. The four planets nearest the Sun are all solid, Jupiter and Saturn are the gas giants, and Uranus and Neptune are called the Ice giants, all four are called giants for good reason. As well as these, there are a few dwarf planets. Pluto, which was once classified as a planet but has since been demoted to Dwarf, is one of them. Plus, there are over 200 moons in our solar system.

All but Mercury and Venus have moons orbiting them, with the giant planets having the lion's share.

To give you some idea of the huge scale of our solar system, if there was a road between Earth and Neptune, the furthest planet away from our Sun, and I drove there non-stop 24 hours a day (never re-fuelling or stopping for the toilet of course!), it would take me 3,852 years to get there at a constant speed of 120mph, much faster than I drive my car to my brother's! To do the same journey to the dwarf planet Pluto would take a mere 6,293 years! So yes, our solar system is a pretty huge place.

In understanding that Earth was pretty big when I scaled its size earlier in this chapter, it is then hard to imagine that Neptune has a mass 17 times the mass of Earth, slightly more massive than its nearest twin Uranus which is 15 times the mass of Earth. If Neptune was hollow, it could contain 60 planet Earths inside it, very much a giant indeed. That said, the true giants of our solar system are the gas giants, Saturn being the smaller of the two and most recognised by its rings. It is truly a giant in every sense, over 700 planet Earths could fit inside Saturn.

The biggest planet in the solar system is Jupiter, an absolute monster, so big that experts say that if it was slightly bigger and ignited, it would become a dwarf star. Jupiter is so big that its mass is more than two and a half times that of all the planets in the solar system combined. More than 1,300 planet Earths could fit inside Jupiter. This constitutes our solar system as we know it today, and all the above orbit our star, known as the Sun.

Our Sun is truly enormous. To give you some idea, you could fit 1.3 million planet Earths inside the Sun. You could even fit 1,000 planet Jupiter's inside the sun, so big, hot and powerful that if you brought a tiny piece of the Sun to

Earth's surface, akin to the size of a teaspoon, it would instantly vaporise everything in a 150 mile radius and create a shock wave that would circle the globe.

Even just within the confines of our own solar system, our planet Earth truly is just a speck of dust, a single pixel in a photo, to say nothing of the insignificance of my journey to my brother's house, and thus far on our journey outwards, we have yet to even leave the solar system.

Moving out to take a look at our galaxy, the Milky Way, we really start to see how insignificant we are. Our galaxy is approximately 100,000 light years across, moving at a constant 186,000 miles per second. It would take far longer than the human species' entire known history just to cross our galaxy, to say nothing of its mass. It is estimated that our galaxy contains between 100 and 400 billion stars, and at least as many planets, if not many times that figure. Some of these stars defy logic in their sheer size, literally dwarfing our Sun.

The biggest star discovered to date is the hypergiant star called UY Scuti, which is located about 5,219 lights years away, so pretty close relative to the size of the entire Milky Way galaxy. The difference between our Earth and our Sun is less than the difference between our Sun and UY Scuti.

UY Scuti is 1,700 times the size of the Sun, and by volume it is so big you could fit 5 billion Suns inside it. I am struggling to get my head around something so big! If you placed UY Scuti where our Sun is, it would consume everything within its own mass as far out as Jupiter. If you were to take a photo of UY Scuti and placed our sun alongside of it, our sun would be so small you could not see it, actually smaller than a single Pixel, that is our sun that could fit 1.3 million planet Earths within it.

This is the biggest Star discovered to date, but there are many others that are only slightly smaller, VY Canis Majoris, Ky Cygni, Betelgeuse, to name but a few stars out of many that would dwarf our own Sun. There are many more at or about this size, all contained within our galaxy.

Moving away from our galaxy, we start to look at other galaxies. Our nearest spiral neighbour is Andromeda, which is twice the size of the Milky Way, so by definition must contain twice the amount of stars and planets. Both the Milky Way and Andromeda are part of a group of galaxies in close proximity, in cosmic terms that is, called a local galaxy group. They are in fact, separated by millions of light years.

Expanding even further, when you get a clustering of local galaxy groups together, they become known as super clusters, each super cluster consisting of many local galaxy groups. At this point the mind boggles as to the sheer vastness of what we have discovered about our universe.

To date we estimate that there are over 2 trillion galaxies in our universe. Mathematicians humbly estimate there could be more stars in our universe than grains of sand on the entire planet Earth, and that of course is only what we have discovered so far. The universe could be infinite, or we may have only discovered less than 1 % of what exists, who knows for sure? It's truly guess work at this stage, but we now know it is enormous beyond our comprehension.

When the Hubble telescope was first put into commission, they decided to point it at an area of space believed to be empty. It was just dark space to the naked eye and to the current telescope's technology. Hubble was pointed to an area of space about one 24-millionth of the whole sky, equivalent to holding a tennis ball 100 metres (approximately 300 feet) away from where you are standing, and that would be the size of area being looked at, tiny and smaller than my finger nail.

The results were breath-taking to say the least. In that tiny minuscule area of the night sky, thought to be empty, we discovered over 10,000 galaxies, not stars, galaxies, so far away were they that the light that they omitted would have been travelling for over 13 billion years, 186,000 miles every second for over three times the entire length of our planet Earth's entire history. Now, times that result by the entirety of the sky above us, and try to imagine just how many galaxies there are, to say nothing of stars and planets within them.

Now my journey to my brother's truly is nothing.

In fact, the huge gas and ice giants that are in our solar system are incalculable and tiny when you get to these cosmic scales, they do not even register. Even our Sun would not be recognised in these cosmic scales, it would not even register as even existing, because it would be so small compared everything on a universal scale! Hard to believe but even UY Scuti, the hypergiant star, would be unrecognisable amongst the 10,000 galaxies discovered in a tiny bit of dark night sky, like it does not even exist. To conclude, our entire solar system would not even register as existing, simply too small, yet I would remind you that it is so vast that driving 24/7 at 120mph, it would still take me over 6,000 years to reach the last known planet, the dwarf planet Pluto, yet in the greater scheme of things, our solar system does not even register as existing.

There could be many thousands of stars just as big or even bigger than UY Scuti confined within those 10,000 galaxies or solar systems that dwarf our own. Simply put, even solar systems and enormous stars are beyond microscopic in the expanse of space, leaving our home planet of Earth so small that it cannot be measured, no form of scale is small enough to recognise Earth in the greater universe, it cannot be classified as a tangible existence, as said, we are beyond microscopic.

This is what we know so far, and I am left to think back at my earlier chapters. 'The Earth was made in seven days for mankind' ... well, what was the rest of this unimaginably vast universe made for?

And do we still truly believe that we are alone in the universe?

I have to be honest, and in going over what I covered in the last chapter, in amongst all that we know exists, and there is true credibility now for the fact there are multiple universes as well, I would be more scared if we were alone and the only life existing, than to find out what we really already know (if we are honest with ourselves), and that is that we are not alone at all and the universe is so vast beyond what we can comprehend, and it is mathematically calculated to be absolutely teeming with life.

Chapter 6: The Modern World

Having covered such diverse subjects as our place in the universe, are we alone, our current technological progression, our miss informed history, I thought I would write a chapter about the modern world, the things I see us getting right and the things I see as us getting wrong.

Off course this is just my interpretation and my thoughts based on what I see and hear, they may not be relevant to you dependant on your financial status, lifestyle, education, age or the country you live in, that said I hope you can understand some of what I am saying and can relate to at least some of it.

It is hard to know where to start, as I have so many things I wish to address, I guess a good place would be the current trend that casts name tagging and frankly hurtful comments towards anyone who does not agree with your opinions or point of view. I do not want to crawl down the rabbit hole to far with certain changes that are consuming the headlines at present as I simply do not know enough detailed information to make real tangible comments on those subjects, so many people seem to be experts already, but I do wish to offer my humble take on these things.

The fact I and so many others do not know enough, and maybe reserve our right to not want to know more than we do, is something that is not considered in general for when these modern day changes are discussed, the pure fact that so many people are not fully aware or educated on such matters, and that causes some of the argument's with the issues I see constantly on T.V or read about, I am not alone in my ignorance of the current way of thinking.

Let's consider racism to start with, what I mean by that is that I am not someone who considers himself racist at all, I am a spiritual man and any form of discrimination to me is wrong and hurtful, and during my lifetime I have seen massive steps towards eradicating racism. I grew up in a world of Golly wogs on jam jars, Alf Garnett on TV, and clear racist undertones on programmes like Rising Damp. Now I do not think for one minute that those involved with those things were racist, it was just the norm back then and accepted. As a child I had no concept of racism, and as for picking on physical differences, I went through most of my school life being called big ears, is that a form of racism? I never saw it that way, it was just good old banter and I gave as good as I got back. That said, I can obviously see now how such things that were accepted back then as norm would in no way fit in today's modern world and are clearly racist in nature, and I and most people would agree.

My confusion lies in the other areas of what I grew up with, things like Eddie Murphy in the early part of his career whilst perfuming live openly referring to himself as a Nigger, like wise Richard Pryer, both black comedians, or huge globally successful Bands like N.W.A, " Niggaz with Attitude" a band that consisted of all black members, movies like "Boys in the Hood", where the characters referred to themselves as Niggers. This was not white characters like Alf Garnett or Rupert Rigsby making racist remarks, this was black people making them about themselves, obviously that left huge confusion to your average person on the street, what is acceptable and what is not? This trend continues right up to date with comedy acts performed by the likes of Chris Rock and Dave Shapiro.

I guess what I am trying to convey, and without trying to target anyone, but would particularly like to address this to the younger generation who do seem to have progressed their understanding far greater than someone of my age, probably due to being better educated on these matters and simply did not grow up in my world as outlined above, just because I hear someone tell a simple joke in the privacy of their own home that may have racial undertones or mock the current gender movement, does not mean they are at heart a racist and they do not sympathise with gender identity. Judge people on their character not some silly joke being told by someone who grew up watching Eddie Murphy and Richard Pryer, which has led to a generation of people who were given a green card to tell such jokes by the very people who are targeted in those jokes.

I grew up in a different world, was educated far differently than what is happening today, spent most of my life never having to have to deal with these changes. Not being racist at heart means I go through my day to day life not recognising different skin colour or people of different ethnicity, I genuinely do not see people that way so racism is not a part of my daily life. It does not mean I cannot see the humour in our differences, including my own with my stick out ears. I have held posts as a manager and awarded jobs to people who are of a different ethnicity to me, gender to me and different skin colour, not because I had to meet a quota, or felt compelled to have even numbers of different people, simply because they were the best person for the role, surely that is what I am meant to do. But would

I be classed as racist if I had all white people in a department even if they were proven to be the best for that role?

For me racism only exists when it is made an issue, now that may not be right, it may not suit everyone's interpretation of how we tackle this issue, but for a man who has lived through what I have, seen the massive changes that have already been put into place, recognises more is needed to be done, but has seen things like the N.W.A dominate our music charts and many other things that have led to confusion on the subject, and is someone who would never condone real acts of racism, but also has much less time left on this planet than I have lived. In an ever changing world that has so many platforms now for this subject to dominate so much, I simply ask for you to give me and anyone like me a break, people who grew up ignorant of these things, we are doing our best to understand this new world but it is a far cry from the one that existed when we grew up and were the young adults you are now.

Just because I do not take the knees at a football game does not mean I am racist, it is because I have bad knees that hurt, maybe consider other things before tagging people, before judging people who have lived through more change than you can imagine and may not understand why it is such a huge issue now and simply might not want to engage what years they have left re training themselves to deal with what was never an issue for them in the first place, because they simply were never racist anyhow.

It is the same for the gender movement, now I apologise if movement is the wrong word, again I have no idea what is, and frankly I do not want to know, I am doing my best to understand the changes that are happening, but do not want to submerse myself fully as I simply do not want to. Does that make me a bad human being? Or worse a bigot,

I do not think so but could very easily be classed that way now through lack of understanding of my point of view, I am expected to understand and respect the gender movement but is that reciprocated?

Some of what is happening now it is confusing to me, some of it is frankly crazy, and again I stress, this really is a new thing to most people, I lived the better part of 45 years on this planet without ever hearing about gender identity.

When my children were conceived we had scans done to confirm the gender of our yet to be born children, there were two choices, boy and girl, both my children grew up in a world of two genders, my son went to the boy scouts and my daughter to the brownies, after she first joined what became the Cub scouts after they reformed and accepted girls, she was the first and only girl at that time, even made it into the local paper, this nearly 20 years ago, how forward thinking was I then? That said it is my understanding that a few more girls have joined the cub scouts since then, but most still choosing to attend the brownies, that is majority thinking.

My daughter did not enjoy it and soon asked to join the Brownies with her fellow girls. My children grew up in a world of separate toilets for boys and girls, a place they felt secure and safe from the opposite sex, blue clothes for my baby boy and pink for my beautiful daughter, with white as a neutral colour for both, boys bikes with cross bars and girls with angled cross bars, transformer movies with my son and Hanna Montana with my daughter, bras, knickers, boxer shorts etc, the list is endless, now all of a sudden and overnight, we have anywhere up to 100 genders, and anyone over the age of 25 is expected to understand it, acknowledge it, and have the correct pronouns. Sorry but it is simply going to take a generation or two for this to be Embedded in our cultures, if that is in fact where it concludes, which is no guarantee, there is so much that needs looking at and addressing first I feel, like men overnight identify as women without the gender reassignment surgery done, stripping naked in a girls changing room in front of a child, likewise a women doing the same in front of a young boy, is that really acceptable? I am asking!

Again I am not saying I do not sympathise, I am not a bigot, I am trying to understand, but when you have lived as long as me without this ever being mentioned, it gives you a WTH moment!

I really am as stated earlier in the book, a live and let live person, the fact that we are all different, think differently, have broad and ranging ideas is why the human race has been so successful, if we all thought the exact same way and had zero diversity in this world I assure you our progression would be far slower and it would be a very boring place to live. As far as I am concerned, people can identify as whatever they like, within reason, I will not shy away from the fact that I heard recently there are now 100 different genders to identify as, that to me is something getting way out of hand and possibly in the realms of attention seeking.

Genuine gender identity is for the choice of the individual, I have no issue with that, we must recognise there is a genuine need for this to be accepted but will take time, while these huge monumental changes are being progressed into society, stop looking to be offended at every turn, if I do not ask you how you identify the moment I meet you, is not me trying to offend you or being bigoted, it is just that I spent a life time starting conversations with a hand shake and name introduction. Maybe instead of being offended perhaps consider introducing yourself in name and how you identify, would that be so hard?

With that all said, I cannot dismiss what is fact, and when I hear someone state that a biological man can become pregnant, again for me it falls into the realms of extremism, regardless of how the person is feeling, the simple fact of the matter is a biological man cannot become pregnant, will not menstruate, I do not need to explain the biology behind that, any medically trained person who states otherwise is practicing an agenda and not working within the realms of reality. There is a void at present between identity and what is and remains inescapable facts. I do think that if we stuck to the facts with gender identity, it would be so much easier for people who have no idea what it is about to maybe accept it and embrace it more, stick with what are real gender issues, because right now I have never met a single person who can readily and off the top of their head name every gender there is out there and what they equate too, even those who readily support it, extremism does not help a genuine cause.

Recognising facts and the truth is not bigotry.

I genuinely think too much enthesis is being placed on appeasing the minority, has any consideration been made to how this effects everyone else who is comfortable with the gender they were born with, that is the huge majority.

How does this work in sports is still causing controversy, how can a biological man who is identifying as a women compete in female events? Running, boxing, swimming, to name but a few sports where I am most confidant that in nearly all cases the biological male identifying as a female would win, and win convincingly. Would that be fair on the rest of the participants who are biologically female? it has been proven without doubt by professional tennis players that biological men will succeed over most biological women at their chosen sport.

My son played Rugby for years, and I watched pretty much every game he ever played in, and I can say again without doubt, my daughter would have not only been run off the park by the guys, but probably got very injured too, that is not me being derogatory towards my daughter, she is an amazing, talented successful women, it is just a reality that men are physically stronger than women, facts again, it is pretty much the same right through the animal kingdom too. By the way that is not the fault of males. For me, so many things are happening on what I consider knee jerk reactions without considering how some things affect everyone else, not just the minority and what are the loudest voices right now, but everyone else, a balance needs to be found to accommodate everyone.

I am most sure, like the current and ongoing issues with racism, and the issues of gender identity are here to stay. I just think those involved have to stop casting such cold hearted profanity's at those of us who simply do not understand it, at those who truly are older and have but a few years left of life, the generation above me, who simply do not want to spend their last precious years learning something so radical to them they will never understand it anyhow. If those involved wish for tolerance then practice tolerance first.

Stop expecting everyone to embrace their point of view, people still reserve the right to their own opinion, contrary to common belief, that is their right, I and many others have our opinion and those involved have theirs, it does not mean any of us are right, there is an old saying "agree to disagree" and at minimum understand that huge changes do not happen straight away, historically they can take a generation or two to come to be accepted as the norm.

Maybe those with the loudest voices that are readily dropping to their knee, or embracing gender identity should not expect radical change overnight but take heart that they are the generation that started something that may change the way the human race operates, but it will take time.

Most people on this planet are neither racist or bigots. Education, tolerance and time is the answer, not name calling and demanding all people think the same way as you, you have an opinion and your entitled to it, it does not mean you are automatically right. Everyone is entitled to their opinion no matter what the core subject is. Be tolerant in your dealings with others, through such combinations as the country they grew up in, their age, their financial status, many other things too, means they are probably coming from a completely different world than the one you grew up in, it does not mean that either of you are nothing less than kind, decent human beings.

I really hope that everyone can understand what I have just written, to be honest they were subjects I was going to stay well away from, for fear of being labelled racist or a bigot, but then when I considered the conversations I have had with many people on these subjects, I realised I am not alone in my thought pattern, so many people feel the same as me, so concluded that I should go with my heart and write what I feel, if I have offended anyone, I humbly apologise, that was never my intention.

Anyway, as serious as the other things I will cover are going to be, hopefully they will take a lighter tone, hopefully!

I want to start now by discussing our political landscape and move onto other subjects. Politics at present is frankly nothing more than a complete joke, laced with lies, scandals, corruption and many other negative undertones.

I can only talk about what is happening in my country, England and to an extent the UK.

I think my biggest gripe is the continual belief by politicians that we the masses are not aware of their actions, or cannot see through their lies, it is about time that all politicians understood that they are public servants holding some of the highest office in the land, particularly the Prime Minister and the cabinet ministers, that their sole purpose is to represent the people of this land to the best of their abilities, they need to be whiter than white. They set the rules so abide by them, personal agendas and personal gains should not be their driving goal, sadly that is all I have seen for many years now.

Ever since the expenses scandal some 15 years ago, which let's be honest was a complete joke, there has been a none stop deluge of lies and scandals.

If that was everyday people in everyday jobs, sackings would have happened and arrests made, as it is, we are expected to believe that certain members were not aware they paid their mortgage on their second home a decade previously, or thought that scatter cushions were an acceptable expense to help undertake their political roles, thousands of pounds a month spent on meal expenses and hospitality, bird baths and moats to name but a tiny amount of things that come to light, and what were they given for such criminal activities, the opportunity to pay it back and carry on as normal.

Even as I write this, Nadhim Zahawi has been sacked for serious breaches of ministerial code over his tax affairs, failing to declare millions of pounds in tax to the HMRC. My god, if that was anyone else, even to the sum of just a few hundred pounds, they would stand to lose much more than just their job, and did he apologise for abusing his position, no, he blamed the press.

The last person to enter parliament with true honest intentions was Guy Fawkes. Even if I do not agree with his plan, murder is never the answer, but at least his intentions were true.

There has never been a time in recent years when there is not one scandal or another happening on the political landscape, and these arrogant fools think it is not noticed, I pray for someone of real integrity and honesty to take up the reigns of prime minister and bring some real class to the role, I just do not see it right now.

On the back of Boris Johnson's lies about party-gate, the fact we have had 2 unelected Prime Ministers last year, one still in office, one leaving due to breath taking incompetence, and frankly a country on its knees right now, I have never known a time like right now where I have felt ashamed to call this place my home.

Even when they are questioned on T.V, it is frankly a pointless waste of time, simply yes/no questions are avoided and politicians go off on tangents without actually answering what they were initially asked. If any politician reads this, understand everyone sees through you, a five year old could answers questions better than you do, you should hang your heads in shame.

The political landscape is a joke, our politicians are a joke, there is another old saying " absolute power corrupts absolutely" well in parliament it clearly has.

Infrastructure collapsing, roads collapsing, NHS on its knees, food banks have evolved from help for the needy to being used by professional people like Nurses and Teachers as the cost of living is so high. A national shortage of teachers, dentists, Doctors etc, you cannot get an appointment to see a GP for love or money. I remember a time when they would come to your home if you were ill. People freezing in their own homes, rising cost of living issues, the list goes on.

Continual neglect and self-serving politics has bought us to our knees, and it will be a long struggle back up to make Britain Great again.

What ever happened to public services, things like the railways, buses, etc, they were there to help the people, all we hear now is get out of your cars or we will charge you the Earth.

It is those making those rules that govern who are threatening us with additional costing for using our cars, but it was also them that took away our only viable alternatives to using said car, the Trains and buses are not a good alternative anymore.

Bus route cuts, bus timetables restricted, train branch lines closed down, those choices were a great help to someone elderly living out in the sticks.

Public services are not profit makers!

My experience with trains these days is to stand for long lengths of time as there are no seats available, and it's getting worse.

Further to this, what ever happened to bus conductors, you will have to be of a certain age to remember a conductor, but as a public service they were essential I would say. There to help people who would struggle to get on and off the bus, lift heavy bags and luggage on board, assist customers with their journey, deal with purchasing of tickets, there to advise on what stop to get off and what the next bus number you would need to make your continued journey, something I did all the time as a child, pester the conductor regarding how to get to town and back.

But once again money was the deciding factor, profit was the driving force, and conductors were done away with, public services are not there to make profit, they are there for the public.

I know two people who were bus drivers and who gave it up very quickly because of stress, rude customers, threats etc, because they are now not only driving the bus but dealing with the public.

Does it not strike you as inherently dangerous to have a bus driver in control over so many lives stressed and raging due to his last interactor with a rude customer. And now the latest crack pot idea is they want to do away with guards on trains, how will that help the blind or disabled people, basically they will not be able to use a train service ever again, all in the name of profit and cost cutting, no longer a service for the people.

And still we are told, don't use your car, continued taxation of roads and cities for people who do use their car, in most cases due to nothing more than having no other alternative, or the latest ridiculous idea, 15 minute city's, the mind baffles as to what our so called leaders actually have going through their heads with these crack pot notions.

I have seen this once Great Nation become a run-down mess.

I am not talking about the British Empire which saw us become and remain the country that has invaded more countries and fought more wars than any other country in history, remember that the next time you are quick to judge other country's actions. England has fought in 72 wars since the 16th century, that's only 400 years ago, a rate of one war every 5 years, yet we present ourselves on the world stage as advocates of the abolition of war, a bit hypocritical don't you agree?

I am in fact referring to the country I grew up in.

I have witnessed through my life this country get asset stripped by a succession of politicians, regardless of which party, until now when we are on our knees. No longer the country that gave the world Televisions, motor racing, trains, cars, and the internet, we have no industry anymore, we create very little for the world.

Industry's like our farmers are crying out for help and financial support from our governments, but none is given, having to stay slaves to the supermarket profiteers who have them tied down to minimal payments for such things as dairy products.

Whatever happened to milkmen, fresh milk, fruit juice, bread, eggs, all sorts of product's delivered to your door daily at the crack of dawn ready for when your house hold woke up. Milk in glass bottles, washed and sent back to the local dairy for re-use, the ultimate expression of recycling, now it's big plastic bottles with Milk that can last for weeks, what the hell are they putting in that milk for that to happen?

And why is it that way now, the unstoppable super market juggernauts that drove local business's out of town so some CO can claim a salary 50 times that of the people on the ground, plus a huge bonus. And the company can turn in profits of billions for their shareholders to enjoy.

And where did the Milkmen end up, and so many other small businesses, oh yeah, unemployed or bankrupt.

So much of what I valued in my youth and as a young adult has gone, and the solutions to these problems seem fruitless and without true conviction to address the issues we have.

My earliest memory of something I feel now was probably wrong to take away, was free milk at school every day for children, this was introduced in 1946 and made available to all children under 18. In these hard times with so many people living on budgetary foods I guess that it would be something that would help with the health of young people. In 1968 free milk was abolished in secondary schools and in 1970 it was ended for all children over the age of 7. A number cruncher decided to deprive the last couple of generations of young people this state funded gesture, a great health bonus for all children. Everyone had the choice to have the readily available milk, regardless of financial standing it was available for free for all children right up until 18 years of age, what kind of cruel cold hearted person took that away for the sake of saving money?

I appreciate that fresh fruit is still available for young children in infant schools, and free meals for low income families, but children are children up until 18 years of age, why are we depriving older children a healthy option every day. As stated earlier, asset stripping everything that was good and great about our country. I would guess if I looked closely, this asset stripping pattern is probably shared in every country, the goal is not to improve life these days, but look to where we can cut back and save money.

Maybe some of those billions made for shareholders could be redirected towards the health of our nation's children?

Local shops run into bankruptcy, the high street is all but gone, certainly local small businesses are gone, where I grew up is frankly an eye sore, boarded up shops, no longer required because the super markets have moved in.

Contrary to common thinking, and I guess we are all guilty of using super markets, but they do have a massive negative impact on community's.

Community spirit is all but gone now, we live in a world of pull up the rope I am alright jack.

I remember as a child us having a local Londis shop, a local butchers, paper shop, a bakers, a shop that did school clothes and wool and cotton etc, a hardware store, chemist, hairdressers and off license, the only two remaining now are the chemist and the hairdressers, all the others are gone, boarded up, I do not recognise the area now, it looks so deprived and run down.

Three things have contributed massively to these changes, all within a 1 mile radius of that local line of shops in 3 different directions are a very large Sainsburys, Tesco and ASDA, how can any small business compete with these huge profiteering monsters.

Hardware stores where you could purchase exactly what you needed, like a small number of screws, stood no chance against the juggernauts that are B&Q, Wickes, Screwfix etc. Where you are forced to buy bulk bags of things like screws and nails when only needing just a few of each, is it really the way forward with our declining resources, it is wasteful capitalism again, not looking after the valuable resources of this planet.

Everywhere I go I hear cry's about the long term future of the planet, yet I see hypocrisy in that surrounding me at all times. One example being the rubbish and litter everywhere, this country is a rubbish tip and is treated as such, take a good look at the sides of the roads the next time you are out, they are covered in litter.

Sorry but what kind of brainless moron chucks their used coffee cup, burger wrapper or coke can straight out of the window out onto what can be beautiful countryside. Were you bought up as a child or thrown up?

Is it really that hard to put it in a bin, they are right next to the pump at the garage when you fuel up, that "I am all right jack" attitude on full display.

Nobody these days takes responsibility for their actions, far easier to make it someone else's problem, just chuck it out the window, someone else will pick it up. If that is you reading this, sorry but you nothing more than a selfish narrow minded idiot and if you think that littering the world will help its long term future, think again.

As outlined in the last chapter, we are just a speck of dust in an endless void, maybe we really should start to look after this place, it is all we have.

So much has been lost in our pursuit of improvement because improvement simply cannot go hand in hand with profiteering and capitalism, it is obvious to me now that making money will always mean a cutback in quality. Has anyone else noticed how small chocolate bars are now compared to when I was young, the miniscule amount of crisps in a bag now, the finest examples of cutting back to make profit, quality of product or service provided be dammed.

Our education system is in disarray, lack of funding, lack of teachers etc, has led to a system that really needs a good overhaul.

I worked for an educational trust once, in a managerial role in Estates management, trusts being the government answer for the upkeep of the crumbling fabric of the buildings, most built in the 1950s, to say nothing of the quality of education given, shifting responsibility again.

And I can state categorically, running our schools or academy's as they are now called as a business with huge centralised teams over seeing each academy, taking massive salary's in the process, is not the way forward.

I have seen first-hand the negative effects of this plan, but it did shift all responsibility from government to private trusts I guess, another winning solution from our elected leaders, don't tackle the problem, make it someone else's, even our leaders are guilty of such actions.

Likewise the continued ideas being thrown out by parliament for improvement in other areas, the latest I heard was for pupils to continue on with maths until they are 18, really, that's going to solve society's issues eh!

Most people would have learnt all they need to learn about maths by the age of 12, by that I mean adding, subtracting, dividing and multiplying, it is pretty much all I have needed and I have had a pretty respectful career so far having held multiple managerial roles, safety critical roles, even successfully having my own small business which I started and evolved from nothing, and the reason I can say this is not the right choice is as most people are not academically gifted.

If you have a gift for maths, my son did, then go ahead and continue, learn all about Algebra, Calculus, Geometry, Trigonometry etc, as an academically gifted person you will probably end up in a job reflective of your talents.

My Son ended up getting a commission in the Royal navy and is currently a naval officer, some of the duty's he undertakes like officer of the watch, requires him to take charge of the movements of the ship from the bridge, I have no doubt his academic gifts in many areas, not just maths, have helped him to undertake those duties. The commission alone is something he would not of even been considered for if not for his A-level results, but academic studies are not for everyone, as you would have read in chapter 1, it certainly was not for me, yet I have had a successful life.

My Daughter is as smart as they come, but like me she is not as academically gifted as my son. That said she is gifted in other areas that my son would undoubtedly and at minimum not have been as successful as my daughter has been, and those gifts have seen her become a senior negotiator for a well-established estates agents while simultaneously starting her own small eyebrow business all on her own, all before her 24th birthday.

Studying maths until she was 18 would have been a complete waste of her time.

We should be looking to cater for the individual not the collective, but there simply is not enough money invested in education to allow for such a personal service, but if we are to make things better again, then we need to invest.

Maybe spending billions on a high speed train that will benefit less than 1% of the population of this country was not needed, maybe that money could have been spent on upgrading our education system, or much needed funds for the NHS, two things, the latter especially, that affect everyone.

Instead of rattling on about long term academic studies, try teaching the majority of the next generation who are not academically gifted some real life skills.

How to budget a home, how the banking system works, so that kids today do not get bogged down with loans and credit cards the moment they are old enough to have one. How to save and have a few penny's put by for a rainy day like my Mum taught me. Even things like how to wire a plug, basic plumbing and building skills, inflate your car tyres, check your oil levels or put fuel in the car, cook a dinner, things that will undoubtedly become relevant in their lives and believe it or not will help them much more than knowing all about Algebra or Calculus, real life skills that are just as important and will without doubt be encountered by everyone at some stage.

I am not trying to dismiss further education or in depth specialist study's, far from it, it certainly helped my son achieve his goals and I have extended family members who are still studying, even one cousin taking a PHD.

Both of my Brothers have a degree, my son will end his navel career having acquired through his further education within his roles and rising rank a couple of degrees, I am all for it and would be in full support.

It is just the amount of people who are academic is very minimal to those who are not, so cater for those that are not as well, by giving them the tools they will need for the life ahead of them, the tools to integrate and be beneficial to society and themselves when they become adults. Not force them to study things that frankly they will struggle to take on board and will really only be of benefit to the life of someone academic, who will undoubtedly end up in a job or position reflective of their academic skillset.

These are thoughts that I would imagine would benefit any educational system.

I will look to wrap this chapter up soon, and I hope some of what I have written about has struck a chord with some of you? I do wish to just cover a few more things first though, one of which is our continued conflicts around the world. Surely we can get past all that by now, history has shown that war is not the answer. Less than a century ago we had world war 2, where the enemies of the so called free world were Germany and Japan, yet here we are now in 2023 and Germany and Japan are now allies, the so called enemy is Russia, who were our allies in world war 2, can you see the madness of it all, and still we have conflict on the European stage.

I am not going to get in the politics of this war, I have my opinion, and other than to say be mindful of propaganda and agenda driven news, my opinion will stay with me.

I think what leaves me bemused is the as said proven fact that wars resolve nothing.

Look at Vietnam, so many lives lost and for what? Recent wars when we invaded Iraq, again so many lives lost on the back of a complete lie, no weapons of mass destruction were ever found, but thousands died or were maimed, the conflicts in Afghanistan, what a waste of time that was and all very recently.

And now we have another war in Europe, and the rhetoric seems to be to fuel the war and to heighten tensions, when am I going to hear the words "seeking peace" from one of our world leaders.

Surely that is what we should all be working towards, not Armageddon, these idiots whether politicians or just a man on the street seem to have no idea what nuclear conflict means. It is an extinction level event, nobody will win and nobody will live to see the results come in, none of it will matter, so please, let's move on from war and find a solution to end this horrific situation.

Seek peace and find a way that the entire human race can pool its resources, end corrupt communism and corrupt democracy and capitalism, and work towards a greater goal for mankind, for the benefit of all mankind, after all, we are all the same really.

It does not matter what country you live in, your nationality etc, we all have the same worry's and concerns and we all love our friends and families, so let's start being nice to each other and not killing each other.

I may not be a strong supporter of religious writings but for sure one thing Jesus did say that resonates is "treat others as you wish to be treated yourself". If we all followed that advise all global issues would be resolved instantly.

Another thing that will help in the interactions of human beings, and that is to slow down, stop chasing unrealistic dreams and ventures, stop comparing yourself to others and trying in vain to keep up, for there will always be people with more than you and there will always be people with less, accept that fact and live your life for you, not everyone else.

Set your own goals and targets, however humble they may be, they will bring you a great sense of purpose when you achieve them, keeping up with the joneses next door will bring nothing positive into your world.

The only results for this continued rushing around and stressing is a less satisfying life and living a wired, unrelaxed stressful existence.

To give you an example, I know people who spend their days at work, by that I mean they work all days in the working week, then do a couple of evenings a week, then work Saturdays and sometimes Sundays as well, and for what? So they can go on a really nice holiday for two weeks a year.

Now the holiday is no doubt nice, but at what cost, the cost is to spend your whole life at work, and one day when you look back at your achievements and what you did with the gift of life? you will say, oh yeah, I spent it working 50 weeks a year for a quality two week holiday.

No thank you, I would rather cut back on the holidays, do something more realistic within my true budget and not the credit induced fake budget so many live within, and have a fulfilling life all year round, not just for two weeks.

If you are spending your whole existence at work right now and justifying it as paying for a brief holiday, or the extras in life, then please, I implore you, find a way to change how you are living, because frankly right now you are not living at all, all you are doing is working, work to live not live to work.

Once you step off the treadmill of life and find a way to slow down, start appreciating the simpler things in life, you will find your quality of life will improve no end, you will naturally slow down.

If we could all achieve this, there would be less road rage, more curtesy in our interactions, people would be able to go about their daily lives like I try to do, calm and accepting that things will happen that don't suit my day, I am far from perfect but I do try to live by these standards. Stay calm, don't start taking your grievances out on everyone else, we are just tiny insignificant beings on a spec of duct in an endless void, does it really matter if you are 5 minutes late now and then?

Stop looking at the world through your phone screen, when things go wrong, turn on the radio, a little classical music works for me, and take in the beauty of the world around you, the fields, the sky, trees etc, don't get all worked up, don't start driving up the back of the car in front of you, it is not their fault you are running late, don't overtake in dangerous places, do you really want to risk your own death to counter being late by 5 minutes, or risk someone else's life.

When you consider our place in the greater universe, which as said previously, is so small there is no calculation that recognises that we even exist, does any of our actions or stressful situations really matter at all? To say nothing of being 5 minutes late, it really is not such a big deal.

A good place to start is to take a nice walk, turn your phone off, and interact with real people, you will feel so

JUST A THOUGHT

much better about yourself and will start to see the world with real clarity.

Start living your life on your own terms, whatever they may be, and not be driven by a doctrine handed down by generations of miss informed people or feeling like you are a failure if your car is a couple of years older than your neighbours, or that they take a better holiday than you do, or have a new kitchen fitted while yours is getting older.

Chances are the car they have is a huge payment each month on their credit card, and they are still paying off last year's holiday, and the reality is that they are at work most of the time so never get to enjoy that new kitchen, your find that in most cases this is nearer the truth.

You can ponder this while sitting relaxed in your garden on a warm Saturday afternoon in the sun with a nice can of beer, oh! while your neighbours are at work.

Chapter 7: Tidying up

This last chapter is really about me going over some things I said I would cover in this book in earlier chapters and failed to do so. Other than what will be a small closing epilogue, this will be my last chapter so wanted to make sure everything was tidied up before finishing.

Something I said earlier was that I would give a small explanation about perception, which is a funny kind of subject as for all we know, we may all perceive this world as something totally different to the next person, how we perceive things is a personal interpretation.

That said I will try and give an easy kind of explanation about perception and how one person can influence someone else's reality.

An easy example would be that two people go to watch the same movie at the cinema, the same day, the same time, both having the exact same experience, the only difference is that one person loved the movie and the other thought it was rubbish.

Now from this point onwards one of the people tells ten of her friends that the movie is great, they have known her a long time so take her word for it that the movie is great, at least seven of them will go and see it based on her perception of the movie, how she interpreted the movie.

The other person will tell his friends that it is Rubbish, they have known him a long time so take his word for it that the movie is rubbish, at least 7 of them will not bother to go and see the movie, based on his perception of the movie, how he interpreted the movie.

This pattern will repeat itself with the ten people who go to see the movie based on the original two peoples differing perception's, seven of them expecting a great film, 3 expecting a rubbish film but wanted to see for themselves.

From this new ten people more diverse and differing perceptions will arise influencing many more people about something they have yet to see for themselves to say nothing of the other ten people who never went at all who may well have gone if not influenced by the original two.

This is a very basic way of show casing how we all perceive things differently, even though they are the exact same thing, but can also spawn huge influence over how other people perceive things, even though they have no root facts to base that perception on as they have never seen the movie at that time.

This is how we ended up thinking the world was flat, that Area 51 did not exist and was the subject of myth, why we thought the sun rotated around the Earth, why we considered UFOs/UAPs to be fictional when in fact they were real.

Simply put, someone of influence either perceived things to be that way and we all followed suit, or they managed to convince us that our own perception of these things was how they wanted us to perceive them.

What is perception other than what we perceive something to be at that moment in time, our perceptions of our reality can change daily, hourly, even just by the minute, what we consider false one minute can become fact based the next, a change in our perception can change our reality.

In chapter one I mentioned that I was stopped by an entity from going too far, and I also mentioned the dark. Now I do not know much about this subject and am not an expert in what is the dark, other than to say the laws of Ying and Yang state that everything has an opposite, up and down, left and right, male and female, and off course light and dark.

I guess this would mean that for every good deed done, for every good person, for every act that works within the spiritual rules of love light and truth, there has to be an opposite.

The above is the symbol of Ying and Yank.

During my life I have as stated worked closely within spirituality, and have on occasion been asked to conduct spiritual clearings, also I have worked in carrying out spiritual investigations. For sure during those activities I have felt and sensed entity's (for want of another word) around me that would mean me harm. Now whether they could actually do me any harm remains a mystery, but for sure I felt very uncomfortable around them and being located where I was, this being the polar opposite of what I have felt when sat in spiritual circles or worked with my guides, which is nothing short of a positive warming experience.

So I can only conclude that the laws of Ying and Yang, which go back thousands of years, and came from some very spiritually enlightened cultures, are right. If there is a light side to everything then there is also a dark side, that could explain all the evil acts in the world that leave us feeling sickened when we here of them.

I mentioned earlier that that there could be some truth to the fact that we are not from this world, possibly we were bought here or possibly were genetically modified by advanced aliens who from the primitive beings already occupying Earth created us.

I say this as it relates to the brief statement I made earlier about humans being about the only living thing on earth not fit for purpose on this planet.

If we in fact evolved here then surely like every other life form, by that I mean anything from a honey bee to the blue whale, we would fit the environment for which we live in, we would have evolved and adapted to fit that environment, yet we have not!

There is not one environment on this planet that we do not need to compensate for with such things as extra clothing in the cold, sun cream and sun glasses in the hot sunny environments, breathing apparatus in water, we need machines to fly etc.

So with understanding that information and recognising that every other living thing does fit its living environment, I must conclude that we did not fully evolve here on Earth, we were either bought here from another planet or genetically made here.

The one stand out part about humans is our vast intelligence compared to the other life forms, again we are vastly different, not just by a small margin, but completely dominate on this planet with regard to intelligence.

Now we know the religious teachings are not holding up so well when set against our scientific discovery's with regard to our possible beginnings.

With this new information it also now appears that evolution may also not be the answer to our beginnings either, we should be living in harmony with the planet, simply put we are not fit for purpose on this planet and at best work against nature with our destructive ways.

One possible answer for this is the writings about the Anunnaki, an alien race from the planet Nibiru. These writings are found on ancient Sumerian tablets.

The Sumerians are considered the creators of civilisation, known for their innovations in language, governance, architecture and more, considered a very advanced race who amongst their many city's also built and occupied what is believed to have been the biggest city on earth at the time with a population in access of 80,000 people.

Nibiru is thought to be the lost ninth planet of our solar system, thought to be on a very large and almost weird orbit around our sun, as such it takes many thousands of years to complete one orbit. The dwarf planet Pluto takes 248 years to complete one orbit, what takes Earth 365 days, so it could be possible for a planet to take so long to complete a very long orbit around its host star, our sun.

Nibiru was written to be the home of the Anunnaki, an advanced alien race.

The Sumerian tablets state that the Anunnaki came to Earth to when Nibiru was within close proximity to Earth, they came here to mine for Gold. The Anunnaki had through their own industrial progression done damage to their planet and needed Gold to repair their depleted atmosphere.

It is stated that the Anunnaki created Humans by genetically modifying what they found on Earth, primitive animals, we were created to be a slave race to mine the gold for them.

The tablets also state that the Anunnaki became fond of humans and taught us all about civilisation, all the things the Sumerians are credited for came from the Anunnaki.

The interesting thing about these writings are the Anunnaki are depicted as being anything for from 15 to 20 feet tall, that gives credibility to giants occupying the Earth in the past.

Gives credibility to the fact we were created in his image, Anunnaki are human in appearance, just bigger.

Gives credibility for why we are so intelligent and the fact we are not fit for purpose with this planets environment, because we did not evolve here, we were made.

If you look at the Apollo landers that went to the moon, the shielding on the outside is made of Gold, the best reflector of solar radiation, which would be needed in outer space, that gives real credibility for the use of gold to replenish a planetary atmosphere, written thousands of years prior to us starting a space programme.

All that and to say nothing about why the fact Gold is considered so valuable here now.

Lastly every civilisation on every continent has reference to a great flood in the past.

Nibiru is thought to be about 9 times the size of the Earth. If we consider the effect that our Moon has on our oceans, literally through the moons gravity it creates all planetary tidal movements, imagine what a planet 9 times the size of Earth passing at close proximity would do to the oceans. The moon is one fourth, or a quarter the size of the earth, I would guess something 9 time bigger than Earth would disrupt our oceans on a biblical scale causing global flooding, just a thought!

Lastly for this chapter I wish to cover the universal conscience.

It is my firm belief that we are all connected to the universal conscience, whatever dimension you reside in, whatever universe you reside in, whatever galaxy, whatever planet, whatever country etc, wherever and whoever you are we are all connected.

For me being a spiritual man and having felt things around me that I could not see with my eyes, have passed information onto people I could never have known about, I must concede that there is so much more to, well just about everything, than we are fully aware off.

Meditation is a way and means to break down the energy levels to be able to engage the universal conscience, there is so much history on this matter if we step outside of the mainstream education system.

Native Indians meditated, they were able to predict the forthcoming winters and make contingency plans to survive, they were very enlightened about how to co-exist with planet earth. Samurai warriors also practiced meditation, they meditated upon death daily so that they could fight without fear of death, knowing full well they would return to the universal conscience. Buddhist Monks meditate and mediate a great deal. They do this to reach a state of enlightenment.

For me enlightenment is to emerge from self-imposed Nonage (The period of a person's immaturity or youth, this can take many lifetimes and reincarnations to achieve fully).

Enlightenment is to have the ability to use one's own understanding without another's guidance, to understand the universe and improve your own condition, to pursue knowledge, to understand the values of human happiness, plus many more things that I am most sure being linked to the universal conscience, that then links us to many other lifeforms, some far more in advance than us and having far greater understanding of just about everything, would help achieve.

These are just a couple of examples of meditating noted as being used through history.

We are all connected, this is how we get inspiration, it comes from the universal conscience, ideas, empathy, telepathy, it is all that connection we share.

The hippies had it right, trees, oceans, planets are also all a part of the conscience, mother earth is called mother earth for a reason, we are all part of the same Conscience that spreads its wings right through every life form and dimension that exists, and that is more than we can currently comprehend.

Now I was going to wrap this up now, but lastly I would like to share some of my experience with working with spirit, I hope you will find them of interest.

Some of the people involved with these accounts have had their permission sought prior to me writing about them, others have not, but in either case and to protect all I have not and will not be using those directly involved names, all the same I hope you enjoy the amazing outcomes and results, I certainly did.

I used to sit in a small spiritual circle every week, I enjoyed this circle as it was small and I love being around like minded people and sharing our experiences with spirit, sharing life in general and the friendship it provides in a confidential environment. I considered my circle companions to be close friends, we shared things that maybe outside of circle we would not and we also had great fun, laughter and humour is a great catalyst to linking with spirit. I also enjoyed working with spirit in a circle environment as it was a very relaxed place to connect to the universal conscience. Sadly we all drifted apart as life does but I look back with fond memories.

On one occasion I recall we had had our meditation as usual, and being a small circle, seven people at full strength but normally only four at a time, we each had our opportunity to give an account of what happened.

For me that was normally always short, sometimes I just felt like I had slept, sometimes I got visions, normally on a universal scale, and sometimes nothing other than just enjoying the relaxed environment. This was opposed to others in the circle who sometimes took grand trips and saw grand visions, but we are all different and all receive and work with spirit in a different way, as said, no pressure to conform, we did it our own way.

On this night we finished meditation and I was asked to give what had happened to me while in this state, and all I had to offer was a relaxed time and that someone had said to me that I was going to get a small windfall, not life changing money but a small windfall, and that when I go away on holiday, it would be the start of a heat wave.

I made a mental note of this as I had learnt after a lifetime of living with spirit that they were never wrong.

I was left curious about these two small offerings from spirit as I do not play the national lottery, so my windfall would not come from there, that said I do play the odd scratch card for fun, maybe that would be my windfall, this on top of the fact the weather had been terrible for so long.

Before going away on said holiday, I was asked to give a meter reading for my Gas and Electric.

It transpired after giving my reading that I was in credit for them jointly to the sum of £103.00, and that due to this my monthly payment for that August would be £6.00 rather than the usual £109.00.

I considered that my small windfall.

Leading up to the week before I went on holiday, a holiday with my daughter, my brother and all his family, and my mum, going camping in the New Forest, the weather had been raining and overcast, a major concern for me on what was to be my first camping type of holiday.

The day we left to drive down the sun burst through and continued to get hotter until we had the hottest day of the year on the Wednesday and continued super-hot for a further two weeks.

These two things, the windfall and heat wave prediction, were given to me on a night when I was feeling tired and not totally committed to what I was doing, but still spirit came through and as usual were bang on right.

We also did a quick round robin of readings when in circle, which is fun to do and good practice for those who did not do regular readings, and we had all given many amazing readings whilst in circle, but one of the key things I gave once while in this circle has been stuck in my head ever since as I had feedback when it came to fruition.

I was doing a reading for one of my female friends in circle and it was for her then partner, and it was as simple as giving the name Robert, and that it would be significant to her partner soon.

At the time she had no idea who it was and how it could relate but she took it with her and it was left like that.

The reason I remember is that a few days later she text me to say that while at a meal the previous night with her partner, his phone rang and upon answering it he was shocked to hear from his old friend who had emigrated to Australia some time before and then explaining to my female friend he said: it's my old pal from Australia Rob!

Another great feedback reading was under different circumstances than I would normally work with, I happened to be around my Brothers best friend's house, which kind of makes him my friend too, oddly enough to undertake some spiritual work, when I suddenly felt the presence of spirit and felt I had his Grandfather with me.

Not knowing his Grandfather had passed I first got that confirmed and that it was his Dads father and gave a brief description of his appearance, which my friend confirmed all I had said.

I then went on to tell my friend about a fire engine, which his Grandfather wished to share with him, sadly he could not take this at that time, which did frustrate me as his Grandfather was only giving me this and I felt it was a strong link.

I asked my friend to ask his parents if perhaps his Grandfather had got him a toy fire truck as a child, looking back I was clutching at straws to make sense of it I think as this was all I had been given but the link, as said, was strong.

I bumped into my friend a few days later and he came running up to my car to tell me he had worked it out.

When he was about ten years old his Grandfather had taken him to spend the day at the local fire station where he not only got a ride out in the front of the fire truck but also made it into the local paper pictured in the fire truck.

We both now know what exactly his Grandfather was trying to remind him off, their special day spent together.

I was once asked to do a private reading for two people, a couple, who had recently come over from Australia to settle locally to me.

I had never met them before and relished the chance to read for people who had come from the other side of the world.

I first read for the lady, and had some fantastic links and had great feedback from her the next day.

The reading for the man that followed was also very good and we were both happy with the results as he took all I said.

That said, there is one particular part of the ladies reading I wish to cover, that involved her Grandfather.

I explained to the lady that I was working with her Grandfather who was in spirit and that it was her Dads father, which she acknowledged he was in spirit.

I then went on to tell her about a huge argument that had taken place between her Dad and her Grandfather many years ago that although was resolved, it was never quite put to bed and always remained just hidden under the surface of their relationship from that time onwards.

Her Grandfather wanted her to convey to her Dad that he, the Grandfather, was in the wrong about the argument and that he was sorry and he was too stubborn to say it whilst he was alive, and acknowledged that her Dad knows that the Grandfather was in the wrong.

I also said her Grandfather wishes to pass on information about a problem he had with his right foot.

After giving this information the lady explained to me that she had never met her Grandfather as he died before she was born, she took the information with her and that was how it was left.

The very next day I had a text from her to say she had spoken with her Dad in Australia, who was reduced to tears as this huge argument had taken place and he always knew his own Father was in the wrong and was very emotional to at last receive this apology from his Dad all be it from a stranger the other side of the world.

Her father also confirmed that his Dad, her Grandfather had suffered sciatica in his right foot for the last twenty years of his life.

Now my younger Brother is not one of those people to totally dismiss what I believe in, I would like to think he does not consider his big Brother a liar, but for sure he sits on the fence, I think he feels that given undisrupted proof he would believe totally, which maybe one day he will get, but in my experience whenever proof is given to sceptics they find a way to dismiss it as they just do not want to believe, it is easier to knock something or dismiss it if we don't understand it, than it is to acknowledge it and learn about it.

That all said where my Brother is concerned, he is a critical care Paramedic and has dealt with and seen death and many other horrors that most of us can't begin to imagine, so I make allowances for his scepticism.

So it was with great shock that I had a call from my Brother one Sunday morning to come down and chat to me about a dream he had the previous night, a Saturday night that involved our Dad, who has passed over.

It was clear to me that my Brother was emotional about this dream, in fact he described it to me as feeling unlike any dream he had ever had, almost akin to an outer body experience, which I knew it was straight away.

During this dream my brother was taken to an end terrace house, although not the same as the house we grew up in, it made sense to me as we both grew up in an end of terrace house with Dad.

From here he was taken on a train to London and then onto Tower Bridge, now apart from feeling very real to my Brother, he was asking me what it was all about.

It was whilst trying to make head or tail of it that I remembered my Son and his girlfriend where going to London on this particular Sunday to watch the Lion King in the west end, so my initial thoughts were maybe it was connected with that, my intuition to link it to my son was proven to be right.

Not knowing my sons plans for the day as they set out from his girlfriend's parents' house, I never got the chance to speak to him until he came home late that night.

It turned out that my son and his girlfriend decided to leave much earlier than they needed too for the show and were on their train to London at the exact time my Brother was relaying this story to me, and that they decided to head to Tower bridge as there was a Christmas market there, where my Son and his girlfriend bought some very expensive cheese for me for Christmas, again all unplanned, this decided at the last minute as Christmas was approaching.

Anyone knowing my Dad knows he loved cheese, in fact a jar that had expensive cheese within it that was my Dads, a present I would imagine, now houses his ashes on my bookcase.

My brother taken on a train to London and tower bridge, telling me about this dream at the exact same time my son was on a train to tower bridge, which none of us knew about and then buying some cheese for which my dad was famous for loving. Coincidences, maybe, but a bloody big ones.

BTW, we all live 65 miles from central London.

EPILOGUE

Firstly, many thanks for reading my book, I really do hope you enjoyed it and hopefully leant some things you may not have considered before.

I just wanted to add as what could possibly be a great way to conclude this book, that things I said about at the beginning of writing this book have now taken some form of fruition.

I wrote that I felt that I was at a turning point in my life where I felt I could possibly be going in another direction.

I have for some time enjoyed the idea of doing some talks about what I write about, having successfully done a talk some years ago on some of these subjects, I wanted to try again.

About half way through writing this book, a project that I fitted into my already busy schedule, I did a talk at my local spiritual church, in front of about 30 people, a night I felt went successfully.

This was arranged very quickly and for me was just a taster exercise to see if I enjoyed it and how it was received by everyone.

I was approached after the talk by a man who sat at the back and whom run's a business involved with professional public speaking and wanted to know if I would be interested in doing some of that work myself. He was interested as he said he had nobody doing talks on the subject material I covered and he himself found the subjects interesting.

Since that time I have now completed this book and am looking to get it published.

I have also written my first professional talk backed up with a power point display, and only last week was contacted by my new friend and now have my first professional and paid talk booked in, the venue in London.

I intend to write and present many other talks on multiple subjects and hope to continue with this path and ultimately showcase all my talks as much as possible and as far reaching geographically as I can..

Could this be that change I referred too and ultimately predicted at the start of this book?

THE END

Milton Keynes UK
Ingram Content Group UK Ltd.
UKHW010653250923
429338UK00001B/14

9 781803 815336